Strange Concepts and the Stories They Make Possible

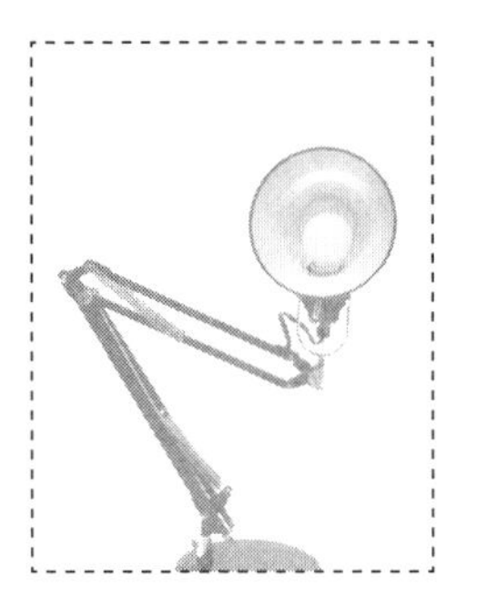

Strange Concepts and the Stories They Make Possible

Cognition, Culture, Narrative

[LISA ZUNSHINE]

The Johns Hopkins University Press
Baltimore

Printed in the United States of America on acid-free paper

2 4 6 8 9 7 5 3 1

The Johns Hopkins University Press
2715 North Charles Street
Baltimore, Maryland 21218-4363
www.press.jhu.edu

Library of Congress Cataloging-in-Publicationa Data

Zunshine, Lisa.
Strange concepts and the stories they make possible : cognition, culture, narrative / Lisa Zunshine.
p. cm.
Includes bibliographical references and index.
ISBN-13: 978-0-8018-8706-2 (hardcover : alk. paper)
ISBN-13: 978-0-8018-8707-9 (pbk. : alk. paper)
ISBN-10: 0-8018-8706-2 (hardcover : alk. paper)
ISBN-10: 0-8018-8707-0 (pbk. : alk. paper)
1. Fantasy fiction—History and criticism. 2. Doubles in literature. 3. Mistaken identity in literature. 4. Robots in literature. 5. Cognitive science. 6. Cognitive psychology. I. Title.
PN3435.Z86 2008
809.3′8766—dc22
2007047748

A catalog record for this book is available from the British Library.

To James Phelan, with gratitude

As far as I am concerned, a mind's arrangement in regard to certain objects is even more important than its regard for certain arrangements of objects, these two kinds of arrangement controlling between them all forms of sensibility.

André Breton, 1928

But who knows what the hell else is going on deep in the soul of a carrot?

Michael Pollan, 2007

[Contents]

Acknowledgments

I am grateful to Susan A. Gelman and James Phelan for their detailed and sympathetic comments on earlier versions of this book. If the final product does not reflect their insightful suggestions, the fault is mine. For the continuous inspiration provided by their remarkable work in the new field of cognitive approaches to literature and culture, I thank Porter Abbott, Frederick Luis Aldama, Mike Austin, William Benzon, Joseph Bizup, Mary Crane, Nancy Easterlin, F. Elizabeth Hart, David Herman, Patrick Colm Hogan, Tony Jackson, Suzanne Keen, Bruce McConachie, Alan Palmer, Alan Richardson, Ellen Spolsky, Uri Margolin, Vernon Shetley, Barbara Maria Stafford, Gabrielle Starr, Simon Stern, Mark Turner, and Blakey Vermeule. At the Johns Hopkins University Press, I would like to thank Michael Lonegro for the care and thoughtfulness with which he approaches projects that cross academic disciplines as well as Juliana McCarthy, Deborah Bors, and Becky Brasington Clark for making the final stages of the work on the manuscript both smooth and enjoyable. I thank Anna Laura Bennett and MJ Devaney for their invaluable editing work at different stages of manuscript production; Jason Flahardy from the University of Kentucky Special Collections for preparing the illustrations; and the College of Arts and Sciences at the University of Kentucky for its support in reproducing the illustrations. Finally, I thank Etel Sverdlov, who makes writing worthwhile.

Strange Concepts and the
Stories They Make Possible

PART ONE

"But what am I, then?"

Chasing Personal Essences across National Literatures

1. *Ural Mountains–Rome–London*

I grew up in a smog-coated industrial town in the Ural Mountains in what used to be the Soviet Union, and now I teach Restoration and eighteenth-century English literature at the University of Kentucky. It is not often that images from my childhood in provincial Russia intersect with images from the world of the late seventeenth- and early eighteenth-century English stage, with their tart political references, powdered wigs, and riot-prone audiences. So when they do intersect, I pay attention.

I was talking with my students about *Amphitryon; or, The Two Sosias* (1690), a brilliant and underappreciated farce by John Dryden (1631–1700), when something about the adventures of one of its title characters, Sosia, made me think of a Russian children's poem that I didn't even know I remembered. I certainly hadn't thought about it for twenty-five years, yet here it was, relentlessly unfolding in my mind and (I felt) interfering with my teaching. For what good would it do to foist on these English majors my Russian-speaking memory? What larger point about Dryden, or *Amphitryon,* or drama, or cross-cultural narrative patterns would it allow me to make? I stifled the recollection and went on with my class.[1]

This happened several years ago, and I have since then figured out why I was so struck by the similarity of the two pieces. The play and the poem both contain the same curious incident (indeed, the poem contains nothing else), in which a character is persuaded that if somebody looks exactly like him, or even just wears his clothing, it must be him. Those persuasion scenes are quite funny, but what I find peculiar is that both authors seem to take for granted a certain philosophical stance on the part of their audience: the assumption that one's true identity (his or her "essence," if you will) goes beyond the sum of that person's attributes and actions.

One may argue that in attributing such a stance to theatergoers in 1690, Dryden was counting on their having been exposed to the relevant ideas of such thinkers as René Descartes (1596–1650) and Wilhelm Leibniz (1646–1716).[2] No such argument, however, can be made about the Russian preschooler whose intellectual horizons hardly extended beyond the politics of the sandbox and an abiding interest in big trucks. In other words, to get the joke of the poem, I had to be much smarter than I was. I was not, and yet I did get it.

I am moving a bit fast here. In a moment I will slow down, go over the poem and the play in detail, and state in more explicit terms why their similarity is so problematic. I just want you to understand why I was so jolted by seeing the two incommensurable cultural environments sharing with such an apparent casualness a complex philosophical assumption. That jolt has made me seek conceptual frameworks that allow for the interplay of the culture specific and the cross-cultural. I have found one such framework in the research of developmental psychologists and cognitive evolutionary anthropologists who study essentialist thinking in young children. Their work has far-reaching implications for all fields of humanistic inquiry, though, of course, I personally am particularly excited by what it means for me and my colleagues in literary criticism who want to understand how works of fiction affect readers.

And to me this research suggests that when we study effects of a given cultural representation, we should inquire, among other things, into the ways it engages our evolved cognitive capacities—the ways it builds on them and experiments with them. In this book, I use this approach to deal primarily with fictional narratives but also with a small selection of movies and surrealist artwork. I hope, however, that as you follow my argument you will think of your own examples of the phenomena that I am discussing, and not just literary, cinematographic, and artistic but also those coming from our everyday interactions with our world.

So let us begin with the Russian poem. Or rather the poem *I* read in Russian but that must have been translated from Polish. Its author, Julian Tuvim (1894–1953), was well known in prewar Poland for his lyrical poems, funny sketches, and political pamphlets. From 1942 to 1946, he lived in the United States, where he published his famous essay "We are Polish Jews." In 1946, he returned to Poland to inter the ashes of his mother, murdered by the Nazis. His lighthearted children's verses were broadly reprinted and anthologized in Russia in the 1970s and 1980s (and still are, as a quick internet search tells me). His ironic aphorisms—"God knows what is going on nowadays! People who had never died before, suddenly started dying"—floated around, delighting readers attuned to this kind of sensibility.[3]

At six years of age I did not know anything about Tuvim, and even if I had, I would not have been able to appreciate it.[4] I must have liked the poem, though, because I immediately learned it by heart. I would like to believe that I actually recall today my feelings on that occasion, and so (if you let me get away with it) I can say that I still remember with what amused superiority I considered its protagonist, a young boy named Yurgán, who wakes up one morning and realizes that he cannot "find Yurgán."

Note that Tuvim does not say that "Yurgán cannot find himself," or something along these lines, because that would complicate his little narrative with the more "adult" overtones of existential angst and difficulties of self-realization, and Tuvim does not seem to be interested in that. His Yurgán quite literally cannot find the physical entity known as Yurgán. He searches the bed, he looks around the house—nothing. He runs into the street where he meets a neighbor, to whom he complains that he "cannot find Yurgán." The neighbor replies, "Look here, you odd fellow, you are wearing Yurgán's jacket! See that torn pocket? You yourself are Yurgán!" The poem ends here with the boy happy about finally locating Yurgán. Thus the apparently absurd initial problem is resolved by winding absurdity to a higher pitch: whoever wears Yurgán's jacket must be Yurgán.

Let's now go back in time even further than late seventeenth-century England—to Rome in 200 B.C. Flush with their victories in the Second Punic War, theatergoers watch Plautus's *Amphitryon,* a tragicomedy that takes the old Greek legend of the conception of Hercules, peppers it with martial allusions, and introduces an important new character, a chatty slave named Sosia. Like Tuvim's protagonist some twenty-one hundred years later, Sosia also comes to believe that he has "lost" himself. But whereas Yurgán checks for his missing self in bed and around the house, Sosia suspects that he left himself "at the harbor"—a grown-up version of Yurgán, whose world has significantly expanded, offering more places in which to "lose" himself but whose reasoning powers seem to have remained peculiarly narrow and thus vulnerable to outlandish suggestions.

Here is how Plautus's Sosia is led to think that he "forgot" himself somewhere else. The Theban general, Amphitryon, has been away from his wife, Alcmena, fighting the army of Teleboeans. The night he is supposed to return home, victorious, Jupiter assumes his shape and seduces Alcmena, impregnating her with Hercules. While Jupiter is busy with Alcmena, he employs another god, Mercury, to keep away Amphitryon's slave—Sosia—who is coming to inform Alcmena about the approach of her husband. Mercury mimics Jupiter's feat of impersonation. He takes on the form of Sosia and tells the real Sosia that *he* is an impostor who has no business hanging around Amphitryon's household. Confounded, the real Sosia begins to mull over the arguments of his assertive double:

> Sosia. God help me, now that I look, he has my features—
> I've seen them in the mirror. We could be twins.
> My hat, my clothes—he's more like me than I am:
> Leg, foot, height, haircut, eyes, nose, even lips—
> Jaws, chin, beard, neck—no difference.

Mercury substantiates his claims to Sosia's identity with heavy beatings, and Sosia is finally persuaded:

> Sosia. Don't, don't—I'll go . . . Gods, help me keep my wits:
> Where was I lost, transformed, made someone else?
> Did I forget and leave me at the harbor?
> He's moved in on my self.[5]

Plautus's play, with its ingenuous doubling of the original twins plot, spawned an "astonishingly long line of . . . dramatic development."[6] Its adaptations include the anonymous English translation of the early 1600s, *The Birthe of Hercules,* Jean Rotrou's (1638), Molière's (1668), Dryden's (1690), Johann Daniel Falk's (1804), and Henrich von Kleist's (1806) respective *Amphitryons* as well as Jean Giraudoux's *Amphithryon 38* (1929), Georg Kaiser's *Amphitryon Doubled* (1944), and others. Sometimes only parts from the original play have been used, such as in Shakespeare's *The Comedy of Errors* (ca. 1594), which is based primarily on Plautus's *Menaechmi* but borrows the scene in which Amphitryon is denied entrance into his own house (because Jupiter is already inside and has been taken for the "real" Amphitryon). Shakespeare portrays Antipholus of Ephesus raging by the barred door of the house he shares with his wife, Adriana, while the twin servants, both named Dromio (and reminiscent of Sosia and Mercury) enthusiastically contribute to the ruckus.

Now we are back in England in 1690. Dryden's play, considered by Earl Miner "one of the unrecognized masterpieces of English comedy,"[7] is based on both Rotrou's and Molière's versions of *Amphitryon,* but it departs from them significantly, expanding, in particular, the theme of the comic predicaments of Sosia—hence the new subtitle, "The Two Sosias." We learn from Judith Milhous and Robert D. Hume's exemplary reconstruction of late seventeenth-century theatrical productions that in the first cast of Dryden's farce the role of Sosia was played by James Nokes. Here is the description of Nokes's typical demeanor in a comic role, provided by Colley Cibber (who joined the Theatre Royal in 1690 and was to become in time a prominent theatrical manager, playwright, and England's poet laureate):

> The lauder the Laugh the graver was his Look upon it; and sure, the ridiculous Solemnity of his Features were [sic] enough to set a whole Bench of Bishops into a Titter. . . . In the ludicrous Distresses which, by the Laws of Comedy, Folly is often involv'd in, he sunk into such a mixture of piteous Pusillanimity and a Consternation so ruefully ridiculous and inconsolable, that when he had shook you to a Fatigue of Laughter it became a moot point whether you ought not to have pity'd him. When he debated any matter by himself, he would shut up his Mouth with a dumb studious Powt, and roll his full Eye into such a vacant Amazement, such a palpable Ignorance of what to think of it, that his silent Perplexity (which would sometimes hold him several Minutes) gave your Imagination as full a Content as the most absurd thing he could say upon it.[8]

We can only imagine with what delicious mixture of "piteous Pusillanimity" and "Consternation" Dryden's Sosia, as played by Nokes, "debated . . . by himself" who he was or was not, all the while surveying surreptitiously the pugnacious claimant to his identity:

> Sosia. He is damnable like me, that's certain. Imprimis: there's the patch upon my nose, with a pox to him. Item: a very foolish face with a long chin at end on't. Item: one pair of shambling legs with two splay feet belonging to them. And, *summa totalis:* from head to foot, all my bodily apparel.

Faced with such compelling evidence, Dryden's Sosia has no choice but to admit sorrowfully that this stranger *is* himself—"there is no denying it." "But what am I, then?" asks the befuddled character as he feels his identity slipping away from him: "For my mind gives me I am somebody still, if I knew but who I were."[9]

What makes the situation funny in all these cases is the naiveté of the protagonists. They are willing to accept that their identities are defined by external characteristics, such as a jacket with a torn pocket, the shape of the "leg" and "foot," "height" and "haircut," or a "patch upon [the] nose" covering a venereal sore. We find the attitude of Tuvim's Yurgán as well as of Plautus's and Dryden's respective Sosias amusing (the latter additionally set off by Nokes's brilliant acting) because it is clear to us that they really *should know better.*

But—and here I pose a seemingly simple question, on whose deceptive simplicity this book's argument will turn—on what grounds should they know better? What is it that I knew as a preschooler, and that the readers of Plautus were aware of two thousand years ago and the Restoration audiences immediately recognized as belonging to the jurisprudence of "the Laws of Comedy," but that Yurgán and Sosia so spectacularly fail to grasp?

To answer this question I need a conceptual framework that can do two things simultaneously. On the one hand, it should allow me to cut across disparate historical settings—for how much really does a child from twentieth-century provincial Russia have in common with an adult theatergoer in second century B.C. Rome or in late seventeenth-century London? On the other hand, it should be sensitive to historical differences, so as not to reduce the cultural meaning of one representation to another.[10] For, clearly, the intended and actual effects (whatever they were) of Sosia's ridiculous reasoning on Dryden's original audience were intertwined with political and aesthetic idiosyncrasies of the 1690s. And, clearly, learning about these effects will explain next to nothing about my reaction to Tuvim's poem or about Plautus's audience's reaction to "their" Sosia—we will need to study those in their specific cultural contexts.

In other words, I need a framework that can handle cognitive similarities and cultural differences and, in the long run, help to explain how the two interact. It is with such requirements in mind that I turn to the framework offered by recent cognitive evolutionary research on children's essentialism, although I invite my readers to disagree with me and come up with an alternative, if only because "my" explanation involves the liability of a many-page theoretical buildup from a discipline relatively unfamiliar to literary critics.[11]

And here it comes.

2. *Essentialism, Functionalism, and Cognitive Psychology*

There are different ways of defining essentialism. Literary critic Diana Fuss calls it "a belief in the real, true essence of things, the invariable and fixed properties which define the 'whatness' of a given entity."[12] Note the term "entity," which implies that, in principle, we can approach anything—from chair to rose and from Dryden to quasar—with a "belief" that there is something "invariable and fixed" about each, something that makes it/him what it is and not something else: the chairness of a chair, the Drydeness of Dryden, the quasarness of a quasar.[13]

Cognitive evolutionary psychologists and anthropologists agree that essentialism is a "belief." That is, it describes the way we perceive various entities in the world and not necessarily the way they really are. At the same time, some of them suggest that there are important differences between our essentializing of artifacts and of natural kinds (such as people, plants, and animals). For example, if I ask you what constitutes the essence of a chair, you will likely bring up the chair's function, saying that we sit on it or that it is made to sit on. In contrast, if I ask you what constitutes the essence of Dryden, we may have a much longer discussion about his appearance, parentage, occupation, and politics and never settle on any one decisive feature or quality, even if we both agree that there was something special about John Dryden that made him John Dryden and not, say, John Milton (another seventeenth-century poet).

In other words, essences of natural kinds seem elusive, in fact, strikingly so when we contrast them with essences of artifacts, which are generally synonymous with their functions. So, to modify Fuss's original definition, we may say that our belief in what constitutes "the invariable and fixed properties which define the 'whatness' of a given entity" is quite different depending on whether this entity is a natural kind or an artifact.

Moreover, it's been suggested that this difference is grounded in the evolutionary history of our species: ascribing functions to artifacts and ungraspable yet enduring essences to plants, animals, and people might have had an adaptive edge in the Pleistocene and as such remained part of our far-from-ideal cognitive makeup. Scott Atran advances this argument in his *Cognitive Foundations of Natural History: Toward an Anthropology of Science* (1990), a study that brings together research in folk taxonomies from around the world, data from developmental psychology, and the history of science and philosophy. I return to the question of the possible evolutionary history of essentialist thinking in the next section; now we shall focus on research in children's cognitive development.

It turns out that already by four years of age children think of natural kinds primarily in terms of their underlying, largely immutable, and invisible essences and of artifacts primarily in terms of their functions.[14] Psychologists view such tendencies as "plausibly innate (but maturing)," that is, environmental input is crucial for them, but it does not fully define their development. For example, it "is not the case that essentialism results from the particular cultural milieu of the typical experimental subject (middle-class, educated, U.S.)"; it is not the case that essentialism "can be deduced by language use"; and it is not the case that children simply "learn" essentialist thinking from their parents. (To see how much of it they may indeed learn from their caretakers, see Susan A. Gelman et al., *Mother-Child Conversations about Gender: Understanding the Acquisition of Essentialist Beliefs* [2004].)[15]

The tendency to essentialize natural kinds does not disappear as children grow up, but rather expands and diversifies its application. In fact, it is this tendency that has allowed people all over the world to develop and maintain for thousands of years complicated systems of folk taxonomies, paradoxically both making possible the advent of contemporary scientific taxonomy and also holding science back. It held science back by exaggerating, in Ernst Mayr's words, the "constancy of taxa and the sharpness of the gaps separating them."[16] That is, we *know* now that species constantly change into one another, but our cognitive tendency to essentialize natural kinds makes realizing that species pointedly *lack* essences an uphill battle all the way. After all, it comes so much easier to us to think along the lines of Rousseau's *Emile* (1762) and imagine the "insurmountable barrier that nature set between the various species, so that they would not be confounded."[17]

The experiments that investigate children's capacity for essentializing often consist of a series of transformations inflicted on an animal or artifact. Since the time of publication of Atran's book, such experiments have been replicated numerous times in a broad variety of cultures, and they have become increasingly probing and complex. In a traditional experimental set-up, a child is presented with a toy animal or a picture of an animal wearing a very convincing "costume" of a different species or with some body parts missing or altered.[18] When asked to comment on the species of a hybrid animal such as a skunk altered to look like a zebra, even three-year-old children judge a skunk to be a skunk. The skunk seems to retain that underlying "skunkness" that makes it different from other animals. A tiger without legs is still a tiger, not a new species of animal. As Paul Bloom observes in *Descartes' Baby* (2004),

> In a child's mind, to be a specific animal is more than to have a certain appearance, it is to have a certain internal structure. It is only when the transformations are described as changing the *innards* of the animals—presumably their essences—that children, like adults, take them as changing the type of animal itself.[19]

To illustrate very briefly how the same belief in essence-conferring innards continues influencing our thinking as we grow up, consider the following: the Soil Association—Britain's leading organization responsible for certifying foods as organic—has a rule that if a certain pesticide (*Bt*) is present within the plant "as the result of genetic modification," the plant is not organic. If, however, the same pesticide is sprayed on the plant with the same goal of protecting it from insect pests, the plant still qualifies as organic. The distinction makes no scientific sense, so it seems that in this case the Soil Association's guidelines reflect primarily our essentialist biases.[20]

Back to children and their insistence that some essential "skunkness" survives the skunk's transformation into a zebra. (Insistence inferred from their responses, that is, not explicitly acknowledged as such: neither children nor adults participating in various transformation experiments speak of essences or are expected to "believe that they know what these essences are.")[21] The implicit belief in essences of animals does not quite translate into a parallel belief in essences of artifacts. A log, for example, is not perceived as having any specific quality of "logness" about it; in fact, it seems to change its "identity" quite often: depending on its current function, it can be perceived as firewood, a bench, or a battering ram.[22] Another artifact, a cup with a sawed-off bottom, becomes a bracelet or a cookie cutter—there seems to be nothing that is perceived as intrinsically "cupful" about it.

The experiments have demonstrated again and again that "transformations that radically change the appearance of an object result in judgments of category change for artifacts but stability for animals, implying that animals—but not artifacts—retain some essential qualities that persist despite external appearance changes."[23] As Bloom puts is (building on the earlier research of Frank Keil),

> [If] one carefully reshapes [a coffee pot], adds and removes parts, so that it can feed birds and looks like a bird feeder, both adults and children would say that it is no longer a coffee pot, it is a bird feeder. . . . Similarly, houses can become churches, computer monitors can become fish tanks, and swords can become plowshares.[24]

There is a debate among cognitive scientists about the *degree* to which we essentialize natural kinds as opposed to artifacts. Atran, Keil, and Gelman would mostly say that "essentialist theory is not usually extended to artifacts." That is, one might talk about the "deep essence" presumably shared by dogs, "but one could not normally study chairs to find out what *really* makes them chairs." In contrast, Bloom proposes that at least on some level, "the mental representation of artifact kinds is quite similar in structure to that of natural kinds."[25] The discussion of such similarities centers on the issue of intentionality, which is not surprising, given that when we start speaking of intentions behind the making of artifacts we are moving over to the domain of strongly essentialized entities—living beings.

Thus, although we do think of artifacts in terms of their functions—and clearly much more so than we do of natural kinds—"having the right function is neither sufficient nor necessary for something to be a member of an artifact kind." As Lance J. Rips has demonstrated, "function is not the sole criterion for classifying artifacts"—just as important are "the intentions of the designer who produced it."[26] For example, say you take an umbrella and refurbish it as a lampshade, adding "a pale pink satin covering to an outside surface, gathering it on top and at the bottom so that it has pleats" and attaching "a satin fringe . . . around the bottom edge" and "a circular frame" inside the top "that at its center holds a light bulb." Then you still use this object to protect you from the rain. What kind of object is it?

In the experiments reported by Rips, the adult subjects classified this new "lampshade" as an umbrella if—and apparently because—it was initially designed as an umbrella. However, specifying the intentions of the object's designer differently "completely changed the way subjects classified the object, even though [it] still resembled members of the alternative category. If [the object is designed] as a lampshade, then it *is* a lampshade despite looking much like an umbrella." In other words, though very much centering on their functions, "the essential properties of artifacts lie in the intentions of their designers."[27]

To further show how intentionality complicates the view that artifacts are perceived primarily in terms of their functions, Bloom makes the two following points. First, sometimes you do not even have to transform the appearance of the artifact to change its category: all you need is a socially recognizable intention to treat the artifact as a member of a different category. Thus, whereas many "artifacts do need to be made from scratch given the sorts of functions they typically fulfill" (say, VCRs), "certain artifacts can be created out of members of different artifact kinds or even natural kinds. Such cases include kinds like *weapon, pawn, toy, paperweight, target, landmark, coin,* and *artwork.*"[28]

Second, when an artifact is damaged and so cannot be used anymore in accordance with its function, it can still be perceived as belonging to its original category if "its *current* appearance and potential use are best explained as resulting from the intention to create a member of [this] artifact kind":

> A clock that needs to be rewound or that does not work because it has been gently hit by a hammer is still viewed as a clock, because the details of its structure are such that the best explanation for how it came into existence is through the intention to create a clock. This is the case even if it can never be repaired. A pile of dust created by hitting a clock very hard with a hammer is not judged to be a clock since its current appearance and use are not best explained in terms of the desire to create a clock. Note that the question is not whether one is *aware* of the intention (by that standard, it would be impossible for something that one knows to have been

> created to be a clock to ever stop being a clock, which is plainly not our intuition); it is whether the *current* status of the entity is consistent with this intention.

So, to bring these two points together, we "infer that broken and unreparable objects can remain members of an artifact kind so long as they are still identifiable as being created with the intention to belong to that kind (as with a broken clock) and that an object can become a member of an artifact category solely through the intention of the creator (as when a penny becomes a pawn)."[29]

I return to our perception of artifacts in section 5 ("Talk to the Door Politely or Tickle It in Exactly the Right Place"), where I introduce the notion of conceptual hybrids—artifacts with lively human personalities. Let me conclude this section by reiterating three precepts that the cognitive scientists whose work I discuss in part 1 hold in common even if they disagree on certain details.

First, to quote Bloom again, "it is unlikely that a single procedure underlies our intuitions about identity for all types of individuals. More plausibly, our intuitions about the identity conditions for entities like chairs [i.e., artifacts] are going to be quite different from our intuitions about people [i.e., natural kinds]."[30]

Second, to turn to Gelman, the claim that we perceive natural kinds in terms of their essences is "not a metaphysical claim about the structure of the world but rather a psychological claim about people's implicit assumptions."[31] What this means is that our tendency to essentialize does not testify to the actual presence of any underlying essences; instead, its cross-cultural prevalence reflects the particularities of the cognitive make-up of our species.[32]

Approached as a psychological rather than metaphysical phenomenon, essentialism emerges as "sketchy and implicit—a belief that category has a core, without knowing what that core is"—as a "placeholder" notion:

> [People may] implicitly assume, for example, that there is some quality that bears have in common that confers category identity and causes their identifiable surface features, and they may use this belief to guide inductive inferences and produce explanations—without being able to identify any feature or trait as the bear essence. This belief can be considered an unarticulated heuristic rather than a detailed, well-worked-out theory.[33]

Hence in the rest of this book, when I use the term "essentialism" (or "essentialist thinking" or "essentializing"), I mean psychological essentialism: a hazy belief rather than well-thought-through theory, which influences our everyday thinking primarily about natural kinds (as contrasted with artifacts) and that can be reinforced or weakened by specific contexts.

Third, the primary conceptual benefit of essentialist thinking is that it allows us to make novel inferences about previously unencountered entities. The real-life consequences of making such inferences, however, are just as often not beneficial at all—they can be harmful, neutral, or so context-dependent that it is impossible to judge their value.

For example, if I encounter a huge animal with sharp teeth, I will not spend much time trying to figure out its intentions. Instead, I will try to run away from it or hide because I intuitively assume that it might be in the nature of this particular specimen (as it is in the nature of some other big animals with sharp teeth) to prey on smaller mammals such as myself. In this particular situation, my quick inferencing might be lifesaving. (It might also be silly, depending on what that animal really is, but it is better not to have taken chances.)

To use a different example, a member of a college job search committee faced with 150 applications for assistant professorship in her department (not an uncommon situation in some humanities programs) may decide to pay more attention to the resumes of people who defended their dissertations at Ivy League schools because she thinks that Ivy League candidates are better on the whole than candidates from other schools. Thus picking up a resume of a random Ivy Leaguer, she feels already armed with some novel inferences about that specific person—she believes that he will work harder, publish faster, and think in a more original way than a random candidate from another pile. (As one scholar put it recently, "It is well known that a person with a PhD or even ABD [all but dissertation] from some programs is far more likely to be interviewed than someone with an equal or even better publication record from other programs.")[34] This essentializing thus may exclude from her consideration some excellent candidates, and it does not guarantee that the chosen candidate will indeed be all that she hoped for. But this is the risk she has chosen to take to save her time and energy.

You can see now how various kinds of stereotyping can build on our essentialist proclivity to make novel inferences about strangers based on limited information about them.[35] We may feel as we engage in such inferencing that we are doing very well—that we are simplifying and making manageable our overwhelmingly complex social world—when, in fact, we are severely limiting our mental and social horizons.

The issue of the relationship between stereotyping and essentializing is extremely complex, and it often hinges on the question of whether or not we tend to essentialize human social groups. It's been suggested, for example, that "children spontaneously explore the social world around them in search of intrinsic human kinds or groups of individuals that are thought to bear some deep and enduring commonality" and that different "cultures inscribe the social environment with different human kinds."[36] In one context, those kinds will be predicated on "common somatic features," in another, "on common occupational destinies," and in yet another, on "sexual differences."[37] Cognitive evolutionary psychologists and anthropologists continue to debate this model; there have been a number of illuminating studies over the last decade, including Lawrence A. Hirschfeld's *Race in the Making: Cognition, Culture, and the Child's Construction of Human Kinds* (1996), Robert O. Kurzban, John Tooby, and Leda Cosmides's "Can Race Be Erased? Coalitional Computation and Social Categorization" (2001), and Rita Astuti, Gregg E. A. Solomon, and Susan Carey's *Constraints on Conceptual Development: A Case Study of the Acquisition of Folkbiological and Folksociological Knowledge in Madagascar* (2004).[38]

Whereas it is beyond the scope of my study to engage with the arguments of cognitive evolutionary psychologists and anthropologists on this subject (I return to the problem of group essences only briefly in part 2, in the section entitled "Made to Pray"), I believe that because these arguments focus on *cognitive construction* of essences, they can provide a powerful interdisciplinary boost for cultural critics examining the workings of our social institutions and ideological formations. For what these studies do is offer a series of crucial insights into the ways we build on our cognitive biases and shortcuts to construct our everyday world, which we then rationalize as "natural."[39]

3. Possible Evolutionary Origins of Essentialist Thinking

Why do we perceive natural kinds in terms of essences? Cognitive psychologists and anthropologists offer several different (though not mutually exclusive) hypotheses about the evolutionary history of essentialism. The disagreement centers around the question of whether essentialism is strongly associated with the domain of folk biology (the view espoused by Atran) or whether it "emerges out of a set of domain-general tendencies" (the view endorsed by Gelman).[40]

Atran suggests that a "cross-cultural predisposition (and possibly innate predisposition) to think [about] the organic world . . . [in terms of underlying essences] is perhaps partly accounted for in evolutionary terms by the empirical adequacy that presumptions of essence afford to human beings in dealing with a local [flora and fauna]." Such a presumption "underpins the taxonomic stability of organic phenomenal types despite variation among individual exemplars."[41] In other words, our Pleistocene ancestors could make certain inferences about every new (previously unencountered) organic specimen if they could recognize it as belonging to a certain category. For example, it would make sense to be wary of any tiger—not just the one that ate your cousin yesterday—because it is in the "nature" of tigers to prey on humans.

Moreover, a tiger with three legs would still be perceived as a tiger, not a new three-legged species of animal with unknown properties, because it is in the "nature" of tigers to have four legs and the exception at hand testifies only to the peculiar personal history of this particular exemplar.[42] Or, to turn this point around, perhaps even the fact that different tigers look different strengthens our intuition that "there is something deeper" that they have in common: something nonobvious that they all share and that "holds them together" as one species.[43]

According to Atran, for something like two million years such essentialist thinking might have served as a cognitive "shortcut," one that was instrumental in helping our ancestors to orient themselves amid the bewildering variety of natural kinds, including poisonous plants and predators.[44] The attribution of imagined essences was useful for categorization and thus contributed to the survival of the human species. As such it was selected for in thousands of consecutive generations and became a part of our permanent cognitive makeup.

In *The Essential Child* (2003), Gelman argues that our tendency to essentialize might be associated with several cognitive domains and not just with the domain of folk biology (which deals with flora and fauna). As she puts it, "the cognitive capacities that give rise to essentialism are a varied assortment of abilities that emerged for other purposes but inevitably converge in essentialism." Gelman lists a series of independent cognitive capacities, such as the "basic capacity for distinguishing appearance from reality," for "causal determinism," and for "tracking identity over time" that "individually and jointly" may lead to essentialism. Viewed as an outcome of several fundamental cognitive processes, "essentialism is something we do neither because it is 'good' for survival, nor because it is 'bad' for people who are manipulated by essentialist rhetoric. Essentialism is something that we as humans cannot help but do."[45]

Note that Atran and Gelman agree on the larger point of seeing essentialism as a "side effect" of other evolved cognitive adaptations. Their difference mainly lies in where they locate the "proper [that is, original] domain" of essentialist thinking.[46] Atran discovers it in folk-biological taxonomies, whereas Gelman finds it in a cluster of cognitive biases not necessarily tied to folk biology.

4. "A bullet's a bullet's a bullet!"

Putting aside the issue of the evolutionary history of essentialism, let us revisit the claim that "all and only living kinds are conceived as physical sorts whose intrinsic 'natures' are presumed, even if unknown."[47] Different cognitive psychologists may subscribe to stronger or weaker versions of this claim, but they all seem to agree with its key implication: the set of inference procedures used to deal with living things must be quite different from that used to deal with artifacts.

But if this is the case, how do we account for the fact that in our everyday lives, we (both children and adults) regularly engage in a broad range of domain-crossing attributions, imputing essences to artifacts (for example, to works of art) and viewing various natural kinds in terms of their functions?[48] Such mental operations are clearly crucial for configuring our relationship with our world. Our language itself is quite sensitive to our domain-crossing tendencies, supplying us with such terms as "objectification," "commodification," "anthropomorphization," "personal teleology," and "fetishization." (Although it also could be argued that the reason we have such terms is that we need some special vocabulary for describing less habitual cognitive operations. After all, we do not use the term "objectification" to describe our perception of an artifact—for *of course* we objectify artifacts.)[49]

But, to repeat the question, if we can objectify living kinds (by attributing functions to them) and essentialize artifacts (by attributing essences to them), why insist on the difference between the inferential processes involved in dealing with living kinds and artifacts?

Gelman's answer to this is that although essentialized artifacts "will support some novel inferences," these inferences are "*quite limited* compared to those of natural kind categories." For example, "artifacts can participate in rich causal theories, including those in archaeology and cultural studies . . . , but such theories concern interactions between the object and the larger world, not properties intrinsic to the artifacts themselves."[50]

To clarify: the inferences that we make about some essential characteristics of a stranger after having placed him into a certain category concern his personal features. For example—to turn to tigers again—I assume that because tigers prey on humans, this *particular* tiger might attack me. To put it in more explicitly essentialist terms (something that we do not do in our everyday reasoning—for this is a hidden, if crucial, conceptual step), there is something ineffable, invisible, that all tigers have in common.[51] That essential "tigerness" makes it very likely that this particular tiger will behave in a certain way when confronted with me (especially if it is hungry).

In contrast, consider an ancient Roman coin unearthed at an archaeological site. We may perceive this coin to be very special. Its long history seems to lend it that ineffable, invisible something that the most meticulously wrought modern copy would not possess. And yet that special something—*this degree of essence*—is not something intrinsic to such coins in general and this coin in particular. Instead it pertains to the coin's history of having participated in the complex social and cultural networks of a long-gone civilization. In the above case of the tiger, we focus on the tiger itself; in the case of the coin, we focus on what the coin represents—the lost social world.

Or think of the first conversation between the nine-year-old Oskar Schell and the hundred-year-old Mr. Black from Jonathan Safran Foer's novel *Extremely Loud and Incredibly Close* (2005). Oskar has been looking through the things that fill the old man's apartment: "the stuff he'd collected during the wars of his life":

> There were books in foreign languages, and little statues, and scrolls with pretty paintings, and Coke cans from around the world, and a bunch of rocks on his fireplace mantel, although all of them were common. One fascinating thing was that each rock had a little piece of paper next to it that said where the rock came from, and when it came from, like, "Normandy, 6/19/44," "Hwach'on Dam, 4/09/51," and "Dallas, 11/12/63." That was so fascinating, but one weird thing was that there were lots of bullets on the mantel, too, and they didn't have little pieces of paper next to them. I asked him how he knew which was which. "A bullet's a bullet's a bullet!" he said. "But isn't a rock a rock?" I asked. He said, "Of course not!" I thought I understood him, but I wasn't positive.[52]

An important context for this strange conversation about rocks and bullets is that Oskar finds Mr. Black "amazing" and is "keeping a list in [his] head of things [he] could do to be more like him."[53] Hence the artifacts that surround Mr. Black are worthy of Oskar's attention because they can be perceived as bearing some mark of the old man's personality: to put it in the terms of our discussion, some of his essence has rubbed off on them.

Think now of Atran's observation that we hierarchize natural kinds but not artifacts and observe how Foer's text makes it possible for us to hierarchize both rocks and bullets and hence to argue about their relative importance in Mr. Black's personal past. We can say, for example, that rocks bear a stronger mark of Mr. Black's personality because they bring back the memories of places (as the narrative seems to imply); or we can disagree with this view and say that bullets provide a more intimate link to Mr. Black because they bring back the memories of life-and-death situations and some of them may have even been extracted from Mr. Black's own body. Hence one can argue that, unlike the rocks, bullets don't even need identifying tags, so special have they become. Again, the fact that such an argument does not strike us as absurd demonstrates that the text has succeeded in essentializing these objects, at least to some degree.

Note though that in the final count there is nothing special about either rocks or bullets—they are important only in so far as they connect Oskar (and us) to Mr. Black. Were Foer so inclined, he could have built his little hierarchical narrative around other items on the fireplace mantel, say those "Coke cans from around the world," showing how each Coke can brings back memories of a specific episode in the old man's eventful life.

Compare this with the case of the excavated Roman coin. There is nothing special about it by itself—it could have been a shard—it is the degree and quality of its relationship with the person that we are interested in (say, an ancient Roman merchant) that imbue it with a unique significance. To repeat: if you essentialize a natural kind, you end up understanding something (whether correctly or not) about the particular specimen in front of you (e.g., a tiger), but if you essentialize an artifact, you end up understanding something not about this artifact (e.g., a bullet) but about the man who keeps this bullet on his mantel (Mr. Black). Hence Gelman's point that the range of inferences you can make about the artifact by essentializing it is much narrower than the range of inferences you can make about, say, a human being, by essentializing him or her.

5. *Talk to the Door Politely or Tickle It in Exactly the Right Place*

Here is another way to illustrate the point that, although we constantly and casually cross the conceptual domains of artifacts and living kinds, the inferences that we can draw from an essentialized artifact are peculiarly limited. Consider a veritable storehouse of such conceptual hybrids: J. K. Rowling's Harry Potter series. At one point in *Harry Potter and the Chamber of Secrets* (1998), a boy named Draco Malfoy flatters one of his teachers, Professor Snape, leaving the latter rather pleased.[54] Even if you have never read any Harry Potter books, you can immediately infer, based on my intentionally stripped-of-any-details account of this episode, that the man named Snape needs to breathe, eat, and sleep; that he can talk on various topics, and that he has a broad range of states of mind; that he may have a special relationship with a group of people that constitutes his family; that he can learn to knit, to play drums, and sauté vegetables; and that he can do all these things not just on Tuesdays, or on special magical days, but every day.

This list may strike you as silly, but the fact that we never even bother to consciously articulate to ourselves these and thousands of other automatically assumed possible qualities of "Snape" only shows how integral—to the point of not even being consciously perceptible anymore—this inferential process is to our making sense of the world. In other words, when we hear that a *human being* is flattered, we use our essentializing capacity to assume, even if not consciously, thousands of things about this person because they are in the "nature" of human beings.[55]

Now think of the doors at Hogwarts, the school of witchcraft and wizardry attended by Harry Potter and his friends. Like the majority of other artifacts in this school, these doors are magical: they "wouldn't open unless you asked politely, or tickled them in exactly the right place."[56] They seem thus as appreciative of a certain form of flattery and attention as Professor Snape. And we clearly have no problems with imagining this strange object that is simultaneously a door and a creature capable of a very human emotion.[57] (It is also important, of course, that by the time we encounter this object we have been prepared to expect various kinds of conceptual hybrids—that is, we already know that it is *that* kind of story.)

Yet compare the range of inferences we can draw from Snape's susceptibility to flattery to that we might draw from the similar susceptibility of the doors at Hogwarts. Can we assume that these doors need to breathe and eat? And if they are capable of appreciating attention while hanging on the door hinges at Hogwarts, will they have the same capability at a different, nonmagical location? Can they learn any of the things that human beings can? Do they feel any particular emotions toward their family? And who is the doors' family, anyway? The people who use them regularly? The people who made them? Other doors? The adjacent walls? Furniture made of the same tree?[58]

The reason that we have a difficult time answering these questions is that we know that it is in the "nature" of doors to provide or debar access to a different space, and that's it. Doors are largely defined by their current and/or intended function, and it is only because we are making some inferences about them that we would automatically make about a flattered human that we may be able to attribute certain thoughts, emotions, and desires to this particular set of doors. In other words, a few essentialism-enabled assumptions about human beings do rub off on the flattered doors, but only a few and even those under some duress.[59]

It is equally significant that no matter how many times we read about a door that appreciates politeness in a novel or no matter how many times we hear, for example, of a crying statue in real life, we will never assume that doors ordinarily can be influenced by flattery or that statues regularly cry. Quite a number of cultural representations thrive—that is, attract our attention and interest—precisely because they suggestively *violate* some sort of important conceptual "boundary" between an artifact and a natural kind, perhaps by forcing on us the inferences that we generally are not prepared to associate with even a somewhat anthropomorphized artifact.

In part 2 I discuss the enduring fascination that we feel toward the ontological hybrids that look and act like human beings and yet have been literally "made," artifact-like, by their creators. We remain, I argue, perennially titillated by robots, cyborgs, and androids because they are brought into the world with a defined "function"—as artifacts usually are—and then rebel against or outgrow that function by seeming to acquire a complex world of human feelings and emotions. What is particularly remarkable about these representations is that they seem to retain their "out-of-the-ordinary" feel no matter how many times we have been exposed to similar conceptual curiosities and no matter how well we have been "prepared" for them by our everyday habit of casual anthropomorphizing.[60]

6. Resisting Essentialism

A broad array of cultural representations and discourses could thus be fruitfully considered as both exploiting and resisting our essentializing proclivities. Since in the rest of this book I will deal primarily with representations that exploit them and experiment with them (i.e., novels, movies, and paintings), let me now touch briefly on discourses that explicitly resist them.

One such discourse is evolutionary theory. There is a long philosophical tradition (going back to Plato and Aristotle) of misinterpreting biology "within an essentialist framework," which means perceiving species as discrete and discontinuous.[61] As Ernst Mayr points out in *Populations, Species, and Evolution* (1970), the "concepts of unchanging essences and of complete discontinuities between every *eidos* (type) and all others make genuine evolutionary thinking well-nigh impossible. . . . [The] essentialist philosophies of Plato and Aristotle are incompatible with evolutionary thinking," which focuses on populations instead of types:[62]

> The assumptions of population thinking are diametrically opposed to those of the typologist. The populationist stresses the uniqueness of everything in the organic

> world. What is true for the human species, that no two individuals are alike, is equally true for all other species of animals and plants. . . . All organisms and organic phenomena are composed of unique features and can be described collectively only in statistical terms. Individuals, or any kind of organic entities, form populations of which we can determine the arithmetic mean and the statistics of variation. Averages are merely statistical abstractions; only the individuals of which the populations are composed have reality. The ultimate conclusions of the population thinker and of the typologist are precisely the opposite. For the typologist, the type (*eidos*) is real and the variation is illusion, while for the populationist the type (average) is an abstraction and only the variation is real. No two ways of looking at nature could be more different.[63]

Mayr further observes that,

> The replacement of typological thinking by population thinking is perhaps the greatest conceptual revolution that has taken place in biology. Many of the basic concepts of the synthetic theory, such as that of natural selection and that of the population, are meaningless for the typologist. Virtually every major controversy in the field of evolution has been between a typologist and a populationist. Even Darwin, who was more responsible than anyone else for the introduction of population thinking into biology, often slipped back into typological thinking, for instance in his discussions on varieties and species.[64]

That Darwin "often slipped back into typological thinking" is particularly striking—and testifies to the powerful grip of essentializing—given that Darwin did more than anybody else to ease the power of that grip on modern science. As Mayr reminds us,

> Virtually all philosophers up to Darwin's time were essentialists. Whether they were realists or idealists, materialists or nominalists, they all saw species of organisms with the eye of an essentialist. They considered species as . . . defined by constant characteristics and sharply separated from one another by bridgeless gaps. The essentialist philosopher William Whewell stated categorically, "Species have a real existence in nature, and a transition from one to another does not exist." . . . For John Stuart Mill, species of organisms are natural kinds, . . . and "kinds are classes between which there is an impassable barrier."

What this means is that "Darwin could have never adopted natural selection as a major theory, even after he had arrived at the principle on a largely empiricist basis, if he had not rejected essentialism."[65] And the fact that even for him this rejection was apparently a continuous conceptual challenge (for otherwise why would he sometimes slip "back into typological thinking"?) makes it less surprising that today we continue to struggle with essentialist "assumptions about category immutability."[66]

For, as Gelman reminds us, on some level, it does feel counterintuitive to imagine that categories "transform from one to another, most notably in evolution."[67] This is why children—and in some cases even adults—have trouble appreciating evolutionary accounts of species origins:

> What appears to be difficult is not the complexity of [the concepts of evolutionary theory], nor the scientific methods underlying the evidence, nor even the technical underpinnings of the work. Rather, even nontechnical concepts such as the following seem almost insurmountable: within-species variability, the lack of any single feature (either morphological or genetic) that is shared by all members of a species, and the lack of biological reality to "racial" groupings of people. These conceptual difficulties call into question whether true conceptual reorganization takes place, or whether instead we are looking at the coexistence of multiple frameworks. . . . Adults remain susceptible to less obvious but still potent essentialist assumptions. In other words, essentialism is not strictly a childhood construction. It is a framework for organizing our knowledge of the world, a framework that persists throughout life.[68]

Gelman's reasoning here raises a series of genuinely difficult issues. First of all, it is easy to misinterpret what she is saying and read into it the assertion that because essentialism "persists through life," its specific harmful consequences, such as gender or racial stereotyping, must also persist through life. Gelman, of course, points out repeatedly throughout her book that "instantiations of essentialism" are "culture specific. . . . Essentialism is a species-general, universal, inevitable mode of thought, . . . but the form that it takes varies specifically according to the culture at hand, with the basic notion of essentialism becoming elaborated in each culture's complex theories of nature and society."[69]

The very fact that essentialism informs quite different cultural formations in every society already indicates that whereas our tendency to essentialize may be "inevitable," there is nothing inevitable or unchangeable about each specific instantiation of essentialism. On the contrary, it seems that by understanding how susceptible we are to essentialist reasoning we can successfully "deconstruct" and demystify each instance of such reasoning and see it for what it is—a specific cultural construction parasitizing on a more general cognitive predisposition.

To add further nuance to the thorny issue of the "inevitability" of essentialism, we may remember that cognitive evolutionary psychologists have argued for some time now that we tend to essentialize abstract concepts themselves.[70] In other words, the very way in which we make sense of such terms as "inevitable" or "inherent" may hinder us from grasping their significance within the evolutionary framework.[71]

For example, there is a huge difference between saying that we may have numerous evolved cognitive predispositions that are indeed "inherent" and automatically assuming, as we often do, that the instantiation of these dispositions somehow defines, delimits, and predicts actual-world outcomes. The latter is completely false, and cognitive evolutionary psychologists are very sensitive to this issue, even though this aspect of their research rarely makes it into the lurid (and heavily essentialist) newspaper accounts of what is or what is not "in our genes."

Philosopher Elizabeth Grosz captures this difference aptly when she calls "Darwin's gift to the humanities" the emerging new understanding that within the evolutionary framework, "being is transformed into becoming, essence is transformed into existence, the past and the present are superceded by the future." She further notes that the "sciences which study evolution—evolutionary biology for example—become irremediably linked to the unpredictable, the nondeterministic, the movement of virtuality rather than the predictable regularity that other sciences tended to seek."[72] It may have taken these sciences "more than two thousand years . . . to escape the paralyzing grip of essentialism," and the escape is far from complete, but it is possible.[73]

The vision of science escaping the grip of essentialism raises a larger issue about the possibility—or even necessity—of the rest of us escaping that grip once and for all. One useful way of approaching this issue is to keep in mind that the cognitive apparatus underlying essentialism is here to stay: "Evolutionary shifts are incredibly slow and gradual. Most psychologists who make evolutionary arguments about human behavior assume that what we see *now* are adaptations to how humans lived hundreds of thousands of years ago (in the Pleistocene)."[74] This is to say that we can certainly gain a better understanding of how essentialism works (and what its evolutionary history might have been), *and* we may consciously confront and systematically eradicate essentialist reasoning in specific areas (e.g., science, social politics, personal prejudices), *and* we can analyze the ways in which works of fiction build on and play with our essentialist biases (as I do in the sections to come)—but while this greater understanding may affect some of our specific behaviors, it will not change the cognitive architecture that underlies essentialism.[75]

Nor is this all bad. Cognitive structures that underlie essentialism are not perfect, but then *none* of our cognition is perfect. Evolution, as John Tooby and Leda Cosmides frequently point out, didn't have a crystal ball.[76] It could not "know" what future challenges the adaptations that evolved to respond to the Pleistocene environments would have to respond to in the future, say, in a developed industrial society.[77] The adaptations that contributed, with statistical reliability, to the survival of human species for hundreds of thousands of years and thus became part of our permanent cognitive makeup profoundly structure our interaction with the world, but, as Gelman observes, they are not perfectly fine-tuned to the world, nor do they have to be:

> It's good for us to be mostly right, but we don't have to be thoroughly, entirely right. We obviously have a lot of reasoning heuristics and biases that work pretty well most of the time, but fail spectacularly in other instances. So I'm not troubled by the argument that we're essentialist even though the world doesn't contain essences. [Moreover], one would expect to see evolutionary shifts only if they add to our inclusive fitness (i.e., reproductive success). Again, it's not clear that abandoning essentialism would (on the whole, overall) be a more successful strategy than embracing it, particularly if one considers the cognitive costs involved in adopting a more refined approach.[78]

A literary-critical study is obviously not the right place to discuss in detail the cognitive costs of the hypothetical "abandoning" of essentialism. Still, specifically as a literary critic, I would like to point out just one aspect of that cost: fiction thrives on our misguided certainty that there is something essential about people that makes them what they are even if we can never quite nail that crucial "something" down. Mismatches, gaps, failures, and misguided certainties are good for our stories, and the sneakier they are, the better the stories might be.[79]

7. *The Ever-Receding "Essence" of Sosia*

> I sense my soul. I know it by sentiment and by thought. Without knowing what its essence is, I know that it exists.
>
> Jean-Jacques Rousseau, *Emile*

I argue shortly that the reason the audiences of Tuvim and Dryden can immediately appreciate the incongruity of their protagonists' overreliance on appearances is that both writers have exploited, not consciously but with complete assurance, our evolved cognitive tendency to think of natural kinds in terms of their invisible and yet enduring essences. First, however, I need to clarify how we move from essentializing natural kinds to essentializing individuals, for these two conceptual operations are not completely identical, and the authors in question certainly rely on the latter.

Commenting on the difference between kind and individual essentialisms, Gelman observes that kind essentialism "takes one crucial step beyond individual essentialism. With kind essentialism the person assumes that the world is carved into preexisting natural categories" (see fig. 1). By contrast, "individual essentialism seems not to require any such commitment to kind realism."[80]

Thus, you may "believe that your beloved pet Fido" has something special about him, "some 'Fido-ness' that he retains over time, that would show up even if he were to morph into a frog or a human, and that he carries with him after death (e.g., into heaven, or on reincarnation)." Importantly, however, this "essence of Fido would be specific to Fido" and *not* something "shared with all other dogs."[81] In other words, the features that (we think) compose the "essence" of each individual specimen are not identical to the features that (we think) compose the "essence" of the natural kind to which this individual belongs.

Hence, as "Lisa Zunshine" I (am perceived to) have that ineffable special something that makes me *me;* as a human being I (am perceived to) have a *different* set of special qualities that aligns me with other specimens of my kind. So, whereas, on the one hand, "essentializing of individual people recruits much the same cognitive mechanisms as essentializing of natural kinds," on the other hand, given a specific context, we can easily differentiate between the two.[82]

Works of literature regularly draw on this ability, as does, for example, Michail Bulgakov's anti-Soviet satire *Dog's Heart* (*Sobachye Serdze*) (1925). The novella centers on a scientific experiment during which a dog named Sharik is surgically transformed into a human being, "Mr. Sharikov," after receiving an implant of a pituitary gland and testicles of a dead criminal. Mr. Sharikov exhibits the felonious inclinations of the deceased owner of the gland and testicles, whose *individual* "essence" has thus survived both physical death and reincarnation into a body of a different species. At the same time, in a darkly hilarious nod to *kind* essentialism, Mr. Sharikov also possesses qualities presumably typical for *any dog,* such as inveterate hatred of cats. (At some point, this budding Soviet proletarian finds employment with the Committee for the Elimination of Stray Cats and is very successful at his job.) We are thus able to make sense of the nuances of Mr. Sharikov's hybrid personality because we are intuitively aware of the distinction between individual and kind essentialism.

And so now we can say that Plautus and Dryden exploit their audiences' tendency to think in terms of *individual essentialism.* What they do is particularly

Figure 1. "Not guilty, because puppies do these things." Charles Barsotti, *The New Yorker,* December 14, 1987. No. 36692.

striking because they truly try to get at those "hidden, nonobvious properties that impart identity," only to demonstrate, ultimately, the futility of any attempt to define such properties.[83] That is, they systematically go through one attribute after another (one's appearance, character traits, family history, personal memories, actions, and social standing), which in principle could—but, importantly, *seem not to*—capture the essence of the individual.

Tuvim does it as well, but in a more modest fashion, keeping in mind perhaps the young age of his intended readers. Yurgán's neighbor uses the torn pocket on his jacket to clinch the argument about the boy's identity ("See that torn pocket? You yourself are Yurgán!"). The appeal to the torn pocket may transcend the purely external—and hence relatively easily dismissible—considerations of identity because the rent garment can, in fact, indicate something about the "inner nature" of Yurgán: perhaps he is sloppy, or rowdy, or absent-minded. Still, the poem's attempt to ground personality in the ownership of a jacket—even if the jacket has come to express some of its bearer's character—registers in our minds as a delightful play, a wink to the reader, who, young as she is, is already expected to know better.

In Plautus's play, Mercury does not merely mimic Sosia's outward appearance but also appropriates his actions and memories. When Sosia attempts to hold on to his identity by first swearing that he is Sosia and then inquiring,

Hasn't our ship just docked from the Port of Persis?
Haven't I come from there with my master's message
Here to our doorstep, my lantern in my hand?[84]

Mercury eagerly responds by elaborating Sosia's story and providing details of Amphitryon's sail from Persis and his earlier battle with the Teleboean king. He even describes a war trophy that Amphitryon keeps in the special chest and intends to give to his wife as a present. Sosia is duly impressed by the stranger's knowledge: "I can't refute evidence. I'll just go and look / For another name for myself."[85] In a last desperate bid for his slipping self, Sosia tests Mercury by asking what was it that he, Sosia, did while Amphitryon was fighting the enemy on the battlefield and Sosia was left all alone in the tent—a "bit of action" that the pretender, as Sosia hopes, will "never be able to review." Mercury immediately replies that he took "a jug of wine and filled up [his own] jug"—the correct answer and the one that compels our poor hero to cry out, "But look: if I'm not Sosia, who am I?"[86]

Dryden draws out the discussion of what Sosia did "all alone" in the tent, mentioning the "lusty gammon of . . . bacon" that Sosia apparently devoured while hiding from the battle, and thus brings into sharper relief the plight of the character who can no longer rely on his personal memories to protect his identity. In doing so, he follows and amplifies Molière's version of the Roman original.[87] He also develops further Molière's suggestion that one's origins and family history appear to be as ineffective gatekeepers of one's self as are private memories. Molière's Mercury overwhelms the dumbfounded Sosia with the knowledge of the intricacies of his family history, as he pontificates:

Sosia is my name, with utter certitude,
The son of Davos, skilled in shepherd arts,
The brother of young Harpax, deceased in foreign parts,
The spouse of Cleanthis, the prude
Whose whims will make me lose my mind.[88]

Dryden's Mercury also recites Sosia's lineage and compounds it with yet another particular that calls on Sosia's intensely personal—and embarrassing—recollection. Speaking of Sosia's wife, he observes that she, with "a devilish shrew of her tongue and a vixen of her hands . . . , keeps [him] to hard duty abed, and beats [him] every morning when [he has] risen from her side without having first . . ."—and here Sosia hastens to interrupt and silence his story by saying that he understands him "by many a sorrowful token." (I take these tokens to be the bruises that Sosia incurs when he attempts to shirk morning sex with his wife.) Sosia then adds in a melancholy aside, "This must be I."[89] He seems to be by now deprived of whatever personal identity could be conferred by genealogy, memory, and sexual history.

The social self is the one to which Sosia continues to cling fast. In Plautus's play, Amphitryon remains Sosia's last hope for refuting Mercury's claims to his identity, although this hope is complicated by Sosia's dreams of freedom from servitude. Having exhausted the "tests" that should reveal Mercury's imposture, Sosia decides to go find Amphitryon:

> Back to the port, now; I must tell my master—
> Unless he doesn't know me—why, by Jove,
> Then I could shave my head for a freedman's cap.[90]

It appears that Sosia's conception of his identity is strongly aligned with his position as Amphitryon's slave. If it so happens that his master does not claim him as his servant, he is truly not Sosia anymore, a proposition that could be loaded with ideological overtones, depending on the historical context of the particular version of the play. For example, in the case of Dryden's *Amphitryon,* Sosia's temporary suspension of his identity—*until he hears from his master*—might be taken as an ironic political comment on the mentality of some of the members of his audience.[91] (Though, knowing how much Dryden liked to provoke without committing himself to any definite political stance, it just as well might not be.)

What is important for my present argument is that, however ideologically suggestive we may find the idea of defining the person's self by his social position, this definition is never *quite* satisfactory either. We may say that different cultures vary in the degree to which they correlate people's personal identity with their social standing. Such correlation might have been significantly stronger, for example, in eighteenth-century England than it is in the twenty-first-century United States. Still, I suspect that even in those cultures in which this correlation is very strong—born of a long tradition and a powerful doctrinal imposition—it is still ridden with cognitive tension that may eventually translate into political action.

And, indeed, when Dryden's Sosia does meet his master, who immediately and familiarly threatens him with beatings—a welcome that according to Sosia's own earlier stipulation should leave no doubt in his mind that he *is* Sosia—it fails to convince him. Sosia readily acquiesces that he is "but a slave" and that Amphitryon is "his master," but his freshly reconfirmed social status does nothing to restore his belief that he is indeed the only true Sosia.[92] The memory of past physical abuse at Mercury's hands, coupled with the evidence of his senses, overrides the testimony provided by the actions of his master. Here is Sosia explaining to the incredulous and angry Amphitryon how he came to think that the other Sosia *is* he and perhaps even better than he:

> Sosia. I could never have believed it myself, if I had not been well-beaten into it. But a cudgel, you know, is a convincing argument in a brawny fist. What shall I say, but that I was compelled at last to acknowledge myself? I found that he was very I, without fraud, cozen, or deceit. Besides, I viewed myself as in the mirror from head to foot. He was handsome, of a noble presence, a charming air, loose and free in all his motions—and saw he was so much I, that I should have reason to be better satisfied with my own person, if his hands had not been a little of the heaviest.[93]

Sosia's reasoning is funny, not least because of his grudging admiration for the other Sosia and his sly intimation that his "absent" self is apparently a gentleman, with "a noble presence" and freedom "in all his motions."[94] It almost seems that Sosia does not mind "sharing" his identity as long as his double's apparently elevated social status reflects well on him.

This scene thus works on many different levels, and the "cognitive" reading that I am offering here does not claim to account for its complicated overall effect on a specific audience. Still, I suspect that *at least one reason* viewers have enjoyed Sosia's ludicrous exchanges with the false Sosia and with his master for the last twenty-two centuries is that these exchanges tease and "work out" our evolved cognitive tendency to essentialize individuals. Plautus, Molière, and Dryden obligingly offer up for our consideration various personal qualities that seem to be able to capture a person's "essence" and then invite the audience to laugh at the naiveté of any character who buys such a reductionist reading of his identity. The laughter, however, barely covers, and is made more poignant by, a certain amount of anxiety.

Here is why. On the one hand, viewers are reminded, as they witness Sosia's misadventures, that one's appearance *certainly* does not define one's identity; one's name *certainly* does not define one's identity; one's social standing does not *quite* define identity; one's memories do not *quite* define identity; one's origins do not *quite* define identity; one's actions do not *quite* define identity—although in different cultural-historical settings, each of these "nondefining qualities" would be weighed differently in relation to the others.

On the other hand, precisely because the "essences" that we attribute to individuals cannot be captured—for thinking that there is an essence is a function of our cognitive makeup rather than a reflection of the actual state of affairs—some nervousness will always accompany any failed endeavor to capture the "core" of the person. Certain "what ifs" will perpetually hover over such endeavors, and one can easily think of cultural discourses and practices that seem to vindicate those "what ifs."

For example, *what if* the person's appearance really expresses something crucial about her core being, and we simply haven't yet found the correct way to map one onto the other? The persistence of various sumptuary laws, such as the English Sumptuary Statutes, enforced from the fourteenth to the seventeenth century, reflects, at least in part, this nagging suspicion. A sub-version of this particular *what if* is the notion that you are what you eat; sumptuary laws, after all, governed not just people's dress but also their diet. Although we may no longer share the following sentiment, we certainly see how in Shakespeare's times people could believe that the "roast beef of Olde England was character-building food, stout-fare for stouthearted men, while it was widely presumed that a vegetable diet made men weak, timorous, and effeminate."[95]

Another example: *what if* one's memory is all that truly differentiates one person from another? Numerous science fiction stories as well as movies such as *Total Recall* play with this idea.

Or, *what if* one's actions "define" one's identity? Think of Saul A. Kripke's argument in *Naming and Necessity* (1980) that if "Hitler had never come to power, Hitler would not have had the property which . . . we use to fix the reference of his name." For Kripke, Hitler's "most important properties" consist in his "murderous political role," just as "Aristotle's most important properties consist in his philosophical work." Of course Kripke immediately adds that "*important* properties of an object need not be essential, unless 'importance' is used as a synonym for essence," but the very fact that he feels compelled to make this distinction shows that this kind of importance can indeed be casually used as a stand-in for essence.[96]

Or, *what if* one's social class truly determines what one is? A broad range of Marxist arguments build on this notion.

Or, *what if* one's origins determine it? To make such an argument today one would use, or rather misuse, research on DNA and genes. But even apart from misinterpreting genetics, the argument that one's parentage gets very close to capturing one's essence can be difficult to refute. Again, Kripke acknowledges this difficulty when he asks whether the Queen of England—"this woman herself"—could "have been born of different parents from the parents from whom she actually came? Could she, let's say, have been the daughter instead of Mr. and Mrs. Truman?" His answer is that, whereas we can imagine numerous scenarios in which "various things in [Elizabeth's] life could have changed, . . . what is harder to imagine is her being born of different parents. It seems to me that anything coming from different origins would not be this [woman]."[97] Mercury thus must have dealt Sosia quite a blow when he told him that *he* is the "son of Davos, skilled in shepherd arts" and the "brother of young Harpax, deceased in foreign parts." If Mercury is *that* father's son and *that* man's brother, then he *must* be Sosia.

Yet another *what if* is a combination of several *what ifs*. One can say, for example, that what makes Sosia Sosia is not just that he looks like Sosia and wears his apparel, that he is the son of Davos and brother of Harpax, and that he is Amphitryon's servant, and that he used his time alone in the tent to steal wine, but that *all* these qualities come together in one individual. To adapt Kripke's argument, the essence of the person would thus be determined "not by a single description but by some cluster or family." Of course, as Kripke points out, such reliance on the "bundle of qualities" is just as illusory: if a quality that we seek to define—such as the essence of a person—is "an abstract object," then "a bundle of qualities is an object of an even higher degree of abstraction."[98] As such it can hardly capture anything.

Particularly if there is nothing to capture. But, because we (cannot help but) assume that some essence *is* there, our failure to capture it—again and again and again—does not invalidate our implicit belief in it. Instead, this failure fosters a continuous uncertainty about what does or does not truly express the "core" of a person, an uncertainty that takes myriad cultural forms. Let us see now how a staged exploration of this uncertainty—particularly when it takes a comic form—can do more than just titillate the audience's essentialist biases. Let us see how it can further engage these biases by presenting viewers with a visual embodiment of a conceptual conundrum—two people who look exactly the same and hence might share the same essence. (Mightn't they?)

8. Identical Twins and Theater

Theater has been obsessed with twins since time immemorial (an obsession now fully shared by every national cinematic tradition that I am aware of). Perhaps we can develop a more nuanced understanding of this enduring fascination by contemplating the possibility that, just like the motif of the mislaid identity, the twin theme engages in a particularly focused way a cluster of cognitive adaptations associated with our essentializing tendencies. As such, the dramatic representation of identical twins may offer a sort of pleasurable cognitive "workout" for these adaptations. This is a highly speculative point, but let us see how far it will take us.

"When a myth resorts to twins," writes René Girard in his essay on Plautus's Amphitryon, Molière's *Amphitryon,* and Shakespeare's *The Comedy of Errors,* "it must be trying to make a point and this point cannot be the difference between the twins, otherwise why resort to twins? The point is obviously the absence of difference."[99] But the "absence of difference" can be the point—that is, can be intensely fascinating for the audience—only if that audience immediately recognizes it as blatantly contradicting certain assumptions about the world. And, indeed, argues Girard,

> In . . . all comedies of twins, the characters deal with each other on the assumption that all of them are unique and immediately identifiable as such. If they did not cling to this assumption in the face of contrary evidence they would not become so confused. It takes more than the presence of undistinguishable twins, I repeat, to generate the comic effects; it takes this persistent refusal to acknowledge the possibility of beings, human or divine, less different from each other than we would like them to be.[100]

Let me restate in stronger terms the point that Girard only hints at: the reason we can immediately appreciate the stubbornness with which the befuddled stage twins "cling to [the] assumption" of their uniqueness is that we share that, however battered, assumption. Note that I am not saying that it is somehow fundamentally wrong to assume that one is different from other people. I am saying rather that our reasoning about our personal uniqueness and our attempts to "define" what makes us different could be profoundly informed, and perhaps "fatally flawed," by our essentialist thinking.

And so Sosia's ridiculous exchange with Mercury replicates and parodies the audience's own implicit search for the evidence of some "core" difference between the two men. Having registered what he perceives as Mercury's striking resemblance to himself, Sosia reaches, one after another, for the conceptual knobs onto any of which we, too—had we been in his situation—would have pegged the evidence of our personal uniqueness (name, parentage, personal history, memories, social standing), and each one of them fails him.

We laugh when Sosia gives up after trying those knobs because we feel that, unlike this poor chump, we do have extra knowledge that allows us to tell the real man from the impostor. *Read: we do have extra knowledge that continues to provide welcome input for our essentialist biases, even when Sosia himself runs out of such material.* First, in our privileged position of spectators, we know that the other Sosia is "really" Mercury because we have seen the first act of the play in which Jupiter asked Mercury to impersonate Sosia.[101] Second, we are abetted in our superior knowledge of the "real" difference between the two Sosias by the material realities of a theatrical production.

The latter point is important and bears historical elaboration. A curious observation emerges out of the history of the production of plays featuring twins, such as *The Comedy of Errors.* On the one hand, as John M. Mercer notes, although its plot is "blatantly nonrealistic . . ., audiences and reviewers of *The Comedy of Errors* . . . at least since the Restoration, have demanded a high degree of realism in the stage portrayal of the twins. In response, adapters, directors, and actors have done their best to satisfy this demand."[102] On the other hand, on the occasions when directors and actors truly succeeded in making the twins look identical, their audiences' response was mixed. It seems that the "absence of difference" can be overdone and turn the viewers off instead of keeping them engaged.

For example, in a 1990 production in Stratford, England, when the same actor played both Dromios, artfully working around the moments when the twins have to appear on stage together, one reviewer, according to Mercer, wrote that the spectators "could not enjoy the play's 'errors' of mistaken identity. He argued that it was not funny when the other characters mistook the twins—because he did the same." Along the same lines, another reviewer praised the 1883 production at the Strand in London precisely because Dromio's identity was "never for a moment in doubt, differing as [the two Dromios did] in voice, manner, and appearance." The actions and costumes of the brothers, points out Mercer, "made them enough alike to be funny" and "no more than this [was] required. Drama by definition must create some illusion of reality, and [this play obligates] productions to establish *some* identity between the actors who play the twins."[103]

The BBC's 1983 *The Comedy of Errors,* directed by James Cellan Jones, features one actor, Roger Daltrey, playing the two Dromios, and one actor, Michael Kitchen, playing the two Antipholi. The film uses a split screen in the last scene, when both sets of twins finally meet face to face. Because in this case the "twins" are indistinguishable, there is a real danger of making the action incomprehensible for the audience. To help us tell the brothers apart, Antipholus of Syracuse is portrayed as funnier, more emotionally flexible, and much kinder to "his" Dromio than is Antipholus of Ephesus, who is relatively rigid in his emotional reactions and habitually abusive toward his servant.

Not surprisingly, Dromio of Syracuse comes across as a more cheerful and self-assured person than his much-beaten twin (fig. 2). Subtle differences in wardrobe (at one point, one Antipholus's collar is turned up while the other's is

Figure 2. Scene from *The Comedy of Errors* (1983), in which we are able to see subtle differences between the two Dromios (both played by Roger Daltrey).

turned down [fig. 3]) and stature (one Antipholus seems to slouch more than the other) help us to keep the brothers apart. Still, all these details notwithstanding, at least one critic has insisted that the movie is "made less interesting [because] the actors don't play each twin as having a different personality, so it's difficult to tell who is who."[104]

Can certain properties of our cognitive propensity for attributing essences to individuals account *at least in part* for the audience's preference for twins who can be told apart without too strenuous an effort? When faced with the implied "absence of difference" between the two people, we immediately start looking for any evidence of dissimilarity between them (by contrast, when we see two identical chairs or cups or umbrellas, we do not have the same impulse to tell them apart and construe each as somehow special—unless, that is, they participate in some social interaction that involves *people*). Why do we do it? This is a seemingly simple question that opens up yet another complex issue—the extent to which our social cognition is entangled with our essentialist biases.

For, on the one hand, as *The Comedy of Errors* reminds us, there is a very

Figure 3. Scene from *The Comedy of Errors* (1983), in which we are able to see subtle differences between the Antipholi (both played by Michael Kitchen).

good practical reason for wanting to be able to tell apart the two people who look alike. One of them may owe you money, or he could be your lawfully wedded husband, whereas the other could be a stranger of no importance (yet!) or credit, who is looking to skip town after getting enough gold from the duped citizens. Profoundly social beings as we are, we keep constant track of the (evolving) position of people around us in relation to various social networks that compose the fabric of our lives. The two people who look the same present a challenge to this social awareness, for they cannot possibly have the same communal weight, history, and responsibility, and so not being able to tell one from another can have a significant and possibly negative impact on our own social position.

For instance, Angelo, the goldsmith from *The Comedy of Errors,* stands to lose his credit with respectable merchants after mistakenly giving the golden chain to the wrong Antipholus, and Dr. Pinch, a schoolmaster, is punished for blundering into the Antipholi affair by getting his beard "singed off with brands of fire" and his hair nicked so that he is made to look "like a fool."[105] The danger of losing one's reputation can acquire sexual overtones: Adriana from *The Comedy of Errors* only has dinner with the man who looks like her husband, but Alcmena from *Amphitryon* actually conceives a child with the pretender!

Still, we cannot say that our historically enduring fascination with the theatrical twin motif can be accounted for by our real-life fear of being duped by twins. It is unlikely that at any point in human history—especially given the formerly high mortality of twins, both natural and induced by infanticide—the problem of telling them apart was so pressing as to leave a long-lasting mark on our cognitive makeup. In other words, it is possible that plays such as *Amphitryon* and *The Comedy of Errors* skillfully manipulate our emotions and draw us in by appealing to a complex mix of our social and perceptual biases. These plays speak to our very real need to keep close tabs on *who is who* and *who is where* in our mulitlevel system of social hierarchies. But in addition to that, the plays use this social aspect as a "flag of convenience" in their endeavor to tease and exploit our essentialist proclivities.[106]

That is, the plays featuring twins give us a "cognitively enjoyable" opportunity to exercise our essentialist biases at the same time that they tacitly assure us of the social value of such an exercise. They offer us a safe setting in which we can hunt for visual dissimilarities between the twins and construe those uncovered dissimilarities as "proof" of our intuition about an "underlying, largely immutable, and invisible" essence of each individual.[107] At the same time, they also leave us with a pleasant vague feeling that we have learned some sort of a potentially useful lesson about keeping track of who is who in the social game. Theatrical representations of twins thus may derive much of their emotional appeal from engaging our essentialist biases—playing with them, teasing them, and validating them—but they delight by pretending to instruct.

I propose, only half-jokingly, that Dryden's new subtitle for *Amphitryon,* his "The *Two* Sosias" as well as the late eighteenth-century "improvements" on the title of *The Comedy of Errors*—such as William Woods's *The Twins, or Which is Which?* (1780) and Monk Lewis's *The Twins; or, Is It He or His Brother?* (1799)—implicitly promised their audiences a cognitive workout for their essentializing capacities.[108] It stands to reason, then, that the truly identical stage twins—when they are portrayed by the same actor who makes no effort to emphasize *some* dissimilarity between the two protagonists—give no input, so to speak, to such capacities. Unless some members of the audience take the incidental differences in the actors' portrayal of the twins and construe them (again, enabled by their own essentialist biases) into evidence of the systematic difference between the "brothers," they may not particularly enjoy the twin motif of the play.

To clarify: I am suggesting that completely identical stage twins may *not* provide the best cognitive workout for the audience, especially if we understand the "best" as "the most challenging."[109] There seems to be a fine line here between "good, because quite challenging" and "not that good, because way too challenging." We like a challenge when there is a possibility that we will rise to it, but perhaps we like it less when it appears insurmountable. So one may argue that audiences would not enjoy the portrayal of truly identical twins as much as they would enjoy the portrayal of somewhat identical twins. To put it differently (and to bring in the notion of dramatic irony), other characters on stage might be duped by the twins' appearance, but we should be able to know better. We should be able to see—with some effort, but not too much effort—that they are different, after all.

Such a manageable conceptual challenge would confirm our imperfect intuitions that all individuals are essentially different—even those who appear so strikingly similar. The "comedies of twins" thus tickle and exercise our essentialist biases without questioning them too sharply or attempting to uproot them. For *that,* we have other, darker representational genres (e.g., existentialist fiction) and other, less comfortable cultural discourses (e.g., evolutionary theory).[110]

One wonders, then, about the broader significance of embodiment for cultural representations that engage our essentialism. For clearly it is one thing to read about identical twins and quite another to watch them on stage (or, much more recently, on screen). At any given historical juncture, any given group of people watching a specific performance of *Amphitryon* or of *The Comedy of Errors* experiences a different pull on their essentialist biases because of the physical presence of these specific actors. When late seventeenth-century theatergoers saw that "mixture of piteous Pusillanimity and a Consternation" animating the face of James Nokes's Sosia as he surveyed his "charming" rival to his identity, they enjoyed that particular display of emotions for many different reasons.[111] *At least one reason,* however, for this spectacle being such a treat was that Nokes's facial expression and demeanor were strikingly different from that of the actor playing Mercury.

The viewers thus could fairly feast on the multiple implications that this embodied difference provided for their endeavor of establishing the real Sosia as "essentially" and incontrovertibly distinct from the pretend Sosia. The fact that the real Sosia himself remained left out of this perceptual loop, unable to reclaim his "core" uniqueness based on the difference between his demeanor and that of his haughty double, must have heightened the dramatic irony of the situation and added yet another layer of pleasure to the cognitive workout offered by the scene.

Which brings me back to the quest, articulated in section 1, for a conceptual framework that can handle cognitive similarities and cultural differences. Let us say that we use cognitive evolutionary research on essentialism to study plays featuring identical twins and that we claim—as I do here—that the enduring fascination with the twin motif in comedy owes, at least in part, to this motif's profitable engagement of our essentialist biases. At first glance, this seems to be a very broad, indeed ahistorical, claim. Yet in pursuing it further, I have to keep asking how the specific features of the specific productions must have worked the viewers' uneasy essentialism, for it is clear that it mattered terribly how and when the actors who played the twins signaled their characters' "essential" differences. A cognitive evolutionary framework thus seems to strongly encourage a historicist inquiry on the part of the literary critic.

9. *How Is Mr. Darcy Different from Colin Firth?*

> "I am sure I'm not Ada," [Alice] said, "for her hair goes in such long ringlets, and mine doesn't go in ringlets at all; and I'm sure I can't be Mabel, for I know all sorts of things, and she, oh! she knows such a very little! Besides, *she*'s she, and *I'm* I, and—oh, dear, how puzzling it all is!"
>
> *Lewis Carroll,* Alice's Adventures in Wonderland *(1865)*

I have focused so far on two recurrent themes that build on and play with our essentialist proclivities—the twin motif and the motif of comically mislaid identity. In the concluding sections of this part, I want to indicate the range of other narrative engagements with these proclivities by considering examples from several novels and an autobiography.

I want to show, in particular, in what endlessly nuanced way writers exploit both our eagerness to fix an essence of a given character and our readiness to admit that we have failed, once more, to capture that essence. Contradictory as these two mental stances may seem, they are in fact complementary given the nature of the phenomena that they reflect. Because we grasp for an entity that represents a function of our evolved cognitive makeup rather than a real thing that exists in the world, both our hope and our consciousness of failure are inevitable. Balancing between the two is the stuff our stories are made of.

And so to my first example, which comes from Helen Fielding's novel *Bridget Jones: The Edge of Reason* (1999). Halfway into the story, the irrepressible Bridget flies to Rome to interview the actor Colin Firth, who played Fitzwilliam Darcy in the BBC's 1996 cult production of Austen's *Pride and Prejudice*. The occasion is a double opportunity for Bridget. If the interview comes off well, this one-time assignment may lead to a "regular interview spot" with the *Independent* (or so she hopes).[112] More important, Bridget is infatuated with Firth in the role of Darcy. This is her chance to get on Firth's radar, and who knows what *that* may lead to?

Bridget has been instructed by a "bossy man" from the *Independent* to talk to Firth about his new movie, *Fever Pitch* (1997).[113] She has been explicitly told *not* to discuss with Firth his role of Mr. Darcy, for the staff of the *Independent* is well aware that, given a chance to talk to Firth, any British woman in her thirties would inevitably turn the conversation to *Pride and Prejudice*. Bridget proceeds accordingly:

BJ: What do you see as the main differences and similarities between the character Paul from *Fever Pitch* and . . . ?
CF: And?
BJ: *(Sheepishly)* Mr. Darcy.
. . .
CF: Mr. Darcy is not an Arsenal supporter.
BJ: No.
CF: He is not a schoolteacher.
BJ. No.
CF: He lived nearly two hundred years ago.
BJ. Yes.
CF: Paul in *Fever Pitch* loves being in a football crowd.
BJ. Yes.
CF: Whereas Mr. Darcy can't even tolerate a country dance. Now. Can we talk about something else that isn't to do with Mr. Darcy?
BJ. Yes.

Bridget and Firth then talk about the future of British movies, Firth's Italian girlfriend, Mr. Darcy, Firth's new movie project, Mr. Darcy, *Fever Pitch*, Firth's Italian girlfriend, and, finally, Mr. Darcy again:

BJ: . . . What was it like with your friends when you started being Mr. Darcy?
CF: There were a lot of jokes about it: growling, "Mr. Darcy" over breakfast

and so on. There was a brief period when they had to work quite hard to hide their knowledge of who I really was and—

BJ: Hide it from who?

CF: Well, from anyone who suspected that perhaps I was like Mr. Darcy.

BJ: But do you think you're not like Mr. Darcy?

CF: I do think I'm not like Mr. Darcy, yes.

BJ: I think you're exactly like Mr. Darcy.

CF: In what way?

BJ: You talk the same way as him.

CF: Oh, do I?

BJ: You look exactly like him, and I, oh, oh . . .

(Protracted crashing noises followed by sounds of struggle)[114]

The interview thus closes with Bridget finally losing the self-control that she has fought hard to maintain for the last forty-five minutes and throwing herself on sexy "Mr. Darcy." The scene is hilarious for many reasons, but note how its jokes and ironies build on our essentialist biases, even though we are not aware of it as we read it.

To begin with, Bridget seems to commit the same fallacy as Sosia and Yurgán by trying to persuade Firth that since he looks and talks "the same way" as Mr. Darcy, he is "exactly like Mr. Darcy," which, given that Mr. Darcy does not exist, is as close as one can come to saying that Firth "is" Mr. Darcy. We find this logic funny for precisely the same reason that we find Sosia's and Yurgán's logic funny: we know better. We know the difference between the real and the fictive, and we know the difference between "is" and "like." We know that, however long a list of features that Colin Firth shares with Mr. Darcy one may compile (and Bridget's list, incidentally, is about as sophisticated as the one that convinces Yurgán), that list will always fall short.

But why is Bridget oblivious of all this? Her brain must be addled by her recent romantic disappointment (involving, of all people, a man called "Mark Darcy"), by too much coffee, nicotine, and alcohol, and by overexposure to pop culture. Such a silly Bridget!—we are delighted with the joke.

Not so fast though. The reason the joke works so well is because it is on us as much as it is on Miss Jones. Or, to put it differently, we are able to recognize it as a joke and appreciate it precisely because it builds on a certain conceptual vulnerability that we all share with Bridget. It just so happens that at this point she is the one who falls victim to this vulnerability (and in such an undignified manner, too).

This vulnerability is bound up with our essentialism, of course. To see how it works, first let us reconstruct the literary pedigree of Bridget's comic fall. To do so, we need to return to the point I made earlier when I suggested that had we—you or I—been put in Sosia's position (that is, faced with a pretender to our identity) we would have pegged the evidence of our personal uniqueness on the same conceptual knobs that Sosia does: name, parentage, personal history, memories, and social standing. Some of these knobs would have seemed sturdier than others (say, parentage and personal history[115]), but ultimately none of them would have worked. *Nothing would have worked*—but what other choices would we have had as we reached out into the physical world to grab what really is our own cognitive bias? Ultimately, we would be left out-argued, and so our last decisive statement about our unique identity would be very similar to the one made by Rousseau about his soul: "I sense my soul. I know it by sentiment and by thought. Without knowing what its essence is, I know that it exists."[116] An obviously irrefutable report of one's emotional intuition, but not what you would call a strong logical or philosophical point.

This is to say that it is good that we are never really put in Sosia's position because what his outlandish adventures make clear is that relying on the list of attributes that you think make you *you* will not win that argument. Ultimately, one has to fall back on one's unarticulated essentialist intuitions: "I know that there is something special that makes me *me*." Of course, no Mercury would ever let you get away with this. Sadistic deities are all philosophers-in-training.

This is what works of fiction do to us, then. They create contexts in which characters are forced to come up with a list of attributes that are supposed to define their essence. These lists are typically ridiculous—from the one in *Amphitryon* to the one in *Alice's Adventures in Wonderland*, quoted in the epigraph to this section—and how can they be anything but ridiculous, given the nature of entity that they purport to capture and describe?

This is, then, the old literary tradition that Fielding works with here. First Colin Firth and then Bridget herself are forced to list features that make people different or the same. Of course, Firth is made to sound ironic when he contrasts his two recent characters: "Darcy is not an Arsenal supporter. . . . He is not a schoolteacher. . . . He lived nearly two hundred years ago. . . . Paul in *Fever Pitch* loves being in a football crowd. . . . Whereas Mr. Darcy can't even tolerate a country dance." Still, ask yourself, what makes this irony possible?[117] Firth is flaunting the superfluousness of his arguments, saying, as it were, "This conversation is silly to begin with because what does it matter that Mr. Darcy is not a schoolteacher like Paul? Even if he were a schoolteacher, he still would have been different from Paul because we know that Mr. Darcy is Mr. Darcy and Paul is Paul—there is something so fundamentally distinct about them that we can list these superficial differences until the cows come home and it would not add any extra information to what we all already know."

In other words, the reason that Firth can afford to be ironic is that he is an essentialist talking to another essentialist in a book written by an essentialist for an audience of essentialists. We just *know* that Paul and Mr. Darcy are different and, in the final count, that's what ought to settle it.

So the underlying "cognitive" joke of it all is that Firth may sound ironic and reasonable and Bridget may sound earnest and unreasonable, but both of them appeal to the same essentialist biases. Only Firth relies on these biases to prove-as-it-were-without-proving that there is something fundamentally different about Mr. Darcy and Paul (and then himself and Mr. Darcy) whereas Bridget relies on them to prove that there is something fundamentally the same about Firth and Mr. Darcy. I have already spelled out Firth's unspoken essentialist argument; let me now spell out Bridget's:

Let us say we list all qualities that we can think of that make Colin Firth different from the personage he plays, Mr. Darcy. So after we are done with every single personal attribute, external and internal, which could be described in any human language, and after we have addressed the questions of "reality" and "fiction," can't it still be said that perhaps deep down (it's always deep, isn't it?—has to be deep, for this is how our essentialism works), there is still something ungraspable, undescribable, elusive, but oh-so-important that these two have in common (that they almost physically share in some mysterious fashion), that if only we could describe that something or bring it out in the open for everybody to see, then everybody would be immediately persuaded and would agree that, yes, this something makes them so profoundly and so uniquely similar that who can say that on some level they are not the same?[118]

You may or may not find this argument persuasive. Note that I am not saying that we actually go around thinking in these terms. I am merely saying that now we can understand why we may find this argument persuasive when it is put before us in this fashion.

The suspicion that *something is there* is what in part sustains the mystique of the movies. Our cognitive predisposition to see living beings in terms of their invisible but enduring essences makes it very difficult—perhaps impossible?—to say that the actor and the character whom he portrays (or the two different characters whom he portrays on different occasions) do not share some essential quality that we have not grasped just yet but may still grasp in the future. (Hope springs eternal for those who look for essences.) And if they do share it, how can we argue that on some very important level they are not the same?

This means that we can never settle decisively on either Bridget's or on Colin Firth's side of the argument. Because both of them tap our essentialism in their absolute claims—Firth: "I do think I'm not like Mr. Darcy, yes." Bridget: "I think you're exactly like Mr. Darcy"—neither can be proven unambiguously wrong.

And then it is up to Fielding's individual readers to decide how well these respective appeals to essentialism serve her characters. I would say that protecting his own as well as Paul's and Mr. Darcy's "essential" differences, Firth comes off as a bit too stuffy (which, ironically, realigns him with Mr. Darcy and thus supports Bridget's view of him). Also, Firth may end up feeling annoyed by Bridget's attack, but, in fact, it could have been much, much worse. After all, the old literary tradition to which the character "Firth" belongs is the tradition according to which the characters who are forced to define or defend their unique identity end up either badly beaten up (Sosia) or cuckolded (Amphitryon) or sent falling down the rabbit hole (Alice).

As to Bridget, who seems to be the butt of the joke in this scene, her essentializing may have actually taken her someplace nice. Because for a short while she can't tell (refuses to tell?) the difference between the real man and the fictional character, she briefly gets to inhabit a dream world, in which she can talk to and has a close (though, admittedly brief) physical contact with the man of her fantasy: Colin Firth—Fitzwilliam Darcy—Colin Darcy—Fitzwilliam Firth—movie star—romantic novel protagonist. Not a bad bargain, on the whole. Essentialism has its moments.

10. *Looking for the Real Mademoiselle*

> The most important thing when we design characters is capturing someone's essence.
>
> *Ricky Nierva, from "Pixar: 20 Years of Animation" (2005–6)*

Vladimir Nabokov's autobiography *Speak, Memory* (first published serially in the *New Yorker* between 1948 and 1950) is a chronicle of his charmed childhood in Russia at the beginning of the twentieth century (which ended with his family's close escape from the Bolsheviks in 1919) and of a transformation of a child into a writer. It grows from literary traditions and personal aesthetics very different from those of Fielding's *Bridget Jones*. Yet, it similarly thrives in the narrative space created by our constant dual readiness to search for personal essences and to fail in that search.

Nabokov writes to snatch personal memories, particularly the transient ones of early childhood, from the abyss of indifferent eternity. Looming large on the brink of that abyss is the figure of the narrator's "old French governess."[119] A whole chapter is devoted to her, and by that chapter's end we have learned quite a bit about the "Mademoiselle." We have seen her crying, laughing, pouting, deceiving herself, interacting with various children and adults. We have been given a chance to imagine what she looked like by day, by night, when she was a young girl, and when she became an old lady. We have heard about her personal tastes and reading preferences, about her "limpid" French, about her small and large disappointments, and about her tendencies to romanticize the past and misinterpret the present. After twenty-something pages filled with anecdotes about her adventures in Russia (a country incomprehensible to her), with impressionistic sketches of her appearance, and with the narrator's musings about her thoughts, we cannot help but feel that we know this woman. And this feeling is precisely what Nabokov strives for in his all-out bid to salvage the *real* Mademoiselle from "fading fast" into fiction (for he has earlier used her features for another short story of his, to describe a fictional childhood "entirely unrelated to [his] own").[120]

And then Nabokov reaches for our essentialist buttons and pushes them and shatters our comfortable illusion of understanding the Mademoiselle. It is not that we discover that the information we have been given about this woman is incorrect; rather we are suddenly made to feel that however much we may know about her, our knowledge will always fall short. Something crucial will be missing—something that we didn't even realize we needed to know until we were told that we will never know it. Here is how this feeling of inadequacy and incompleteness is conveyed, in the next-to-last paragraph of the chapter:

> She had spent all her life in feeling miserable; this misery was her native element; its fluctuations, its varying depths, alone gave her the impression of moving and living. What bothers me is that a sense of misery, and nothing else, *is not enough to make a permanent soul.* My enormous and morose Mademoiselle is all right on earth but impossible in eternity. Have I really salvaged her from fiction? Just before the rhythm I hear falters and fades, I catch myself wondering whether, during the years I knew her, I had not kept utterly missing something in her *that was far more she* than her chins or her ways or even her French—something perhaps akin to that last glimpse of her, to the radiant deceit she had used in order to have me depart pleased

> with my own kindness [a reference by the grown-up narrator visiting the aging Mademoiselle in her native Switzerland], or to that swan whose agony was so much closer to artistic truth than a drooping dancer's pale arms; something, in short, that I could appreciate only after the things and beings that I had most loved in the security of my childhood had been turned to ashes or shot through the heart.[121]

As is always the case with Nabokov's prose, so much is going on in this paragraph that I cannot hope to ever unpack all of it with my cognitive-psychological toolbox. But let us proceed slowly and carefully, doing one thing at a time, retracing our steps several times if needed, and see how far it will take us. We begin with the narrator's worry that "a sense of misery, and nothing else, *is not enough to make a permanent soul.*" Our notion of soul has a particular relationship with our essentialist proclivities (I touched on this earlier, when speaking about Rousseau's insistence that he "senses his soul," and I discuss it in some detail again in part 2). It functions in many cultures as an expression of our essentialist intuitions: the "soul" of a person is roughly the "essence" of that person (artifacts do not have "souls" unless they are strongly anthropomorphized or considered in conjunction with a human agent closely associated with them, in which case some of her "soul" or "essence" rubs off, so to speak, onto the artifact).

The narrator thus suspects that misery alone, however compellingly represented, cannot capture Mademoiselle's essence. A cognitive literary critic would agree with this and add that *nothing* can capture a person's essence since our search for that essence and our belief in it are artifacts of our cognitive makeup. But let us bracket that for a moment and move to the narrator's fear that he utterly missed "something in [Mademoiselle] *that was far more she* than her chins or her ways or even her French." Again, we can see now, from the point of view of cognitive analysis, why such a sentiment immediately strikes a chord in the reader. Up until this moment we have been feeling that we were getting to know the Mademoiselle, that the wealth of details, musings, and associations was adding up to an increasingly intimate knowledge of her inner self, a kind of knowledge that admittedly cannot be compressed into a series of simple attributes but still can be conveyed to another person by the same long, impressionistic series of details and musings. And yet note how easy it is for Nabokov to make us feel that he is wrong (and that we are wrong) and that had he been writing (and we reading) about the Mademoiselle not for twenty but for two hundred pages or for two thousand pages, he (we) would have still missed her essence, or as Nabokov calls it on this occasion, "something in her that was far more she."

Broadly speaking, this is the same cognitive dynamic that underlies any situation in which one fictional character tells another (or in which your wife, your husband, your child, your mother tells you) that "after twenty years of living next to me, listening to me, doing things with me, arguing with me, loving me, you still don't know me," and the other has to admit to herself, however reluctantly, that it might be true. The nagging suspicion of not *really* knowing a person—not having grasped that person's "essence"—can be activated in us so reliably because ours is an essentializing species in a world where there are no essences.

But this is broadly speaking. The paragraph that I quoted in full above, however, plays with our ever-hopeful search for essences in a very peculiar fashion. It moves in an almost closed circle of baits and switches: now affirming that we cannot get at the person's essence, now trying to get at it again. (This is not to say, by the way, that Nabokov thinks consciously in terms of essences or the lack thereof, nor that his readers need to assume that there are no essences in the world in order to appreciate what he does. The fact that the paragraph plays with our cognitive instabilities does not imply that we have to be *aware* of these instabilities in order to make sense of what we are reading and enjoy it. Observe here, too, an interesting difference between cultural practices that *have to* be aware of essentialism and should try to transcend it in order to thrive and those that can play with essentialism without any special awareness of it. Thus, science could not proceed beyond a certain point without explicitly confronting essentialism and seeking to escape its grip—as in the case of evolutionary theory—whereas literature can continue to build on and experiment with essentialism without ever knowing that it does so.)

So to return to the issue of the bait-and-switch game (or is it hide-and-seek?) played by *Speak, Memory,* let us reconsider the phrase, "a sense of misery, and nothing else, is not enough to make a permanent soul." From a cognitive evolutionary perspective, this is one sly statement. On the one hand, we have already established that "a sense of misery" cannot capture the person's essence either alone or in cahoots with hundreds of other "senses" because, essentially speaking, there is nothing to capture. On the other hand, "misery" is a loaded term to use in announcing that the essence-hunting project is doomed. "Misery" is an abstract concept, as is power, equality, love, perseverance, ignominy, truth, happiness, and so forth.[122] I pointed out earlier that, for reasons still debated by cognitive evolutionary scientists, we tend to essentialize such abstract concepts just as we do natural kinds. Consider the following brief illustration:

If I ask you about the essence of a chair, you will either automatically adjust the terms and define the chair in terms of its function—for it is an artifact!—and say, for example, "It is made to sit on," or, if it is a particular chair, a favorite of your elderly aunt's, some of what you perceive as her personal essence will rub off on that chair, and you will say, "It is a special chair, my aunt really likes it because . . ." If I ask you (or myself) to define the essence of Lisa Zunshine, we will be talking for a long time, listing attributes and admitting that, yes, all these attributes are correct, but there is still something special that we haven't grasped yet (for example, some peculiarity in the way those attributes are combined) that makes me me and that even my clone will not possess.[123] (Or will it? Clones are so titillating to us precisely because of our tendency to essentialize living beings.)

Finally, if I ask you to define the essence of "power," the conversation will last just as long, if not longer. We will speak of different kinds of power and of certain contexts that completely invert this or that meaning of the term. (For example, if the essence of power is the ability to make other people do one's bidding, is a newborn infant, whose cry sends his mother running across the house, more "powerful" than the president of the country? Is a television program that glues its devoted viewers to their screens more "powerful" than a hurricane that rages outside their houses and makes it advisable to turn off all electrical appliances?) We will resort to different authorities to illustrate our points and positions. You will quote Weber, Foucault, Nietzsche, Gramsci, Freud, Deleuze, and Buddha; I will bring in an example of my cat. I doubt that we will ever agree on any one definition—we will have to put a conscious stop to our discussion by deciding that we have wandered off the topic, or by "agreeing to disagree," or by saying that power is certainly definable but it so happens that we don't have time right now to thrash it all out. Note, too, that this is a rather innocent, "scholarly" version of a disagreement about the nature of power; in everyday life, the fact that power cannot be defined (because we essentialize it) regularly leads to all kinds of political and ideological excesses.

Same with "misery." It is an abstract concept whose essence we sort of intuitively feel but may find difficult to define. And as such it is a very peculiar concept to evoke in a sentence that admits the impossibility of capturing a person's essence. When one entity that we tend to essentialize (a human being) is defined in terms of another entity that we tend to essentialize (an abstract concept), and then the author throws up his hands and says, "no, it really does not work," he has actually put himself into a nice rhetorical position of eating his cake and having it at the same time. The Mademoiselle's essence cannot be defined through her "misery and nothing else," because it cannot be defined through *anything*. Fair enough. Misery, however, is an abstract concept with its own suggestive baggage of uncapturable essences, and as such it is elusive enough to tease the readers forever with the possibility that it *may* actually get close enough to capturing the essence of a person. Thus you may feel that you have indeed grasped something "essential" about a person when you say that she is "all misery," or "all love," or "all vengeance," however little information about that person it actually conveys. (Incidentally, this is how allegories work. For example, in Edmund Spenser's romantic epic *The Faerie Queene* [1589], the young women who dwell in the "House of Holiness," Fidelia, Speranza, and Charissa, embody faith, hope, and charity. They are decisively one-dimensional, but such is the price that the genre pays for forcefully defining people in terms of their single essential qualities. To put it differently, allegory is a cognitive experiment. The genre creates a conceptual framework within which the uncapturable *is* captured, and the narrative assimilates the cost: flatness, stiltedness of characters.)[124]

Moreover, in *Speak, Memory,* the sentence that doubts the effectiveness of "a sense of misery" for making "a permanent soul" sets up a particular pattern of cognitive teasing to be followed for the rest of the paragraph. For note how in the part about "missing something in her" the narrator first announces his failure to capture the Mademoiselle's essence and then immediately tosses us yet another richly suggestive abstract concept that may, or may not, or may, or may not (the teasing never stops) capture that essential "she" of his governess. Let me quote that sentence again:

> Just before the rhythm I hear falters and fades, I catch myself wondering whether, during the years I knew her, I had not kept utterly missing something in her that was far more she than her chins or her ways or even her French—something perhaps akin to that last glimpse of her, to the radiant deceit she had used in order to have me depart pleased with my own kindness.

Thinking back to our experience with Yurgán, and Sosia, and Bridget Jones, we know that there certainly is "something" in the Mademoiselle that is "far more she than her chins or her ways or even her French," and that this "something" cannot be pinned down. But the second part of the sentence—the evocation of "the radiant deceit" that Mademoiselle used during the last meeting with the now adult Nabokov to make him feel good about himself—hints that perhaps that "something" can be grasped, after all. We remember from the early parts of the chapter that the Mademoiselle frequently misjudged her social environment and/or deceived herself about what was going on around her. It seems now, however, that her proclivity for self-deception might have been an expression of her natural tendency to be kind to others in *their* self-deceptions, which indeed means that the narrator has been misinterpreting and underestimating her all along. For if this helpless kindness is indeed her essential quality, it certainly did *not* come through at all in the preceding twenty pages.

And it is only fitting, of course, that we arrive at the intuition of the Mademoiselle's "essential" kindness through the image of the "radiant deceit." Deceit is an abstract concept, and we have already seen that using an abstract concept to express the person's essence has the effect of making us feel that perhaps—just perhaps—we are on to something. The evocation of radiance—another abstract concept with its own complex essentialized qualities—adds new depths to this perception.[125]

But we are *not* on to something! (Or are we?) The project of grasping and holding tight a person's essence is doomed from the beginning, for there is nothing to grasp and hold tight. Hinting that something *is* there, that it *can* be captured, even if the narrator, partially blinded by the "security of [his] childhood" to the true "she" of the Mademoiselle, has missed it once as a child and a second time as an adult while writing the first twenty pages of this chapter works only because it all remains on the level of hinting. Had the narrator said directly that he now understands that the essential thing about the Mademoiselle was that she was helplessly, childishly kind, it would have immediately provoked a contrary reaction in his readers. We would have felt that this is not it, that there is more to it and that the picture of the Mademoiselle's soul is more complex than this, especially after everything we have learned about her in the last twenty pages.

I am moving in circles here and doing it on purpose. I am trying to replicate through the fluctuations of my argument the restless to-and-fro movement inside the conceptual circle in which Nabokov (with quite a bit of help from our evolved cognitive predispositions) locks his reader. For the cumulative effect of that long penultimate paragraph—an effect achieved through the evocation of abstract concepts, on the one hand, and continual affirmations of the failure of the attempt to make a "soul," on the other—is the permanent state of reaching and failing, of hoping to have captured it and of having your hopes disappointed, of learning to hold your intuition lightly, not pressing it too tightly lest it slip through your fingers again. Perhaps when memory does speak, what she says is, "yes, but . . ."

11. *"Mahatma Gandhi: war!" "But he was a pacifist." "Right! War!"*

Foer's *Extremely Loud and Incredibly Close* is a novel about living with one's losses, which is not the same as learning to live with one's losses. When Oskar, a boy who shapes his life around the tragic death of his father, first encounters Mr. Black, he learns that the older man keeps a large biographical index with a card for everyone that he, a former journalist, "ever wrote about," talked to, read books about, came across "in the footnotes of those books"—in other words, "everyone that seemed biographically significant." Each person in this catalogue gets a one-word description, for example, the card for Manuel Escobar reads, "Manuel Escobar: unionist." The brevity of this system initially strikes Oskar as incongruous. Escobar, he points out to Mr. Black, is "also probably a husband, or dad, or Beatles fan, or jogger, or who knows what else." To this Mr. Black (who is deaf and speaks very loudly, hence the exclamation points punctuating his replies) shouts, "Sure! You could write a book about Manuel Escobar! And that would leave things out, too! You could write ten books! You could never stop writing!"[126]

We feel intuitively the truth of Mr. Black's observation; and now, armed with the insights from cognitive evolutionary psychology, we also know why we do. Since we have a tendency to essentialize people, we may indeed write ten books about a person and still be amenable to a suggestion that something "essential" about our subject has been left out. In fact, Mr. Black's catalogue comes to feel both unsettling and compelling precisely because it builds on and teases our essentialist tendencies. To see how it works, let us follow more of Mr. Black's biographical entries as he reads them out loud to Oskar:

"Henry Kissinger: war!
"Ornette Coleman: music!
"Che Guevara: war!
"Jeff Bezos: money!
"Mahatma Gandhi: war!"
"But he was a pacifist," I said.
"Right! War!
"Arthur Ashe: tennis!
"Tom Cruise: money!
"Elie Wiesel: war!
"Arnold Schwarzenegger: war! . . .
"Susan Sontag: thought! . . .
"Pope John Paul II: war!"[127]

Note first of all that the majority of words used to define these famous people are abstract concepts: war, money, music, thought, and so forth. The effect is similar to that achieved in Nabokov's *Speak, Memory,* in which the narrator contemplates "misery" as the defining element of the Mademoiselle's "soul": it works and does not work at the same time. Because our quirky cognitive architecture prompts us to essentialize abstract entities, it is difficult for us not to see "some" truth to almost any pairing of an abstract entity and a person. Consider one of the least apparent pairings: "Pope John Paul II: war." Among the multitudes of meanings, examples, and contexts that we activate when we try to get to the essence of "war," at least one may correspond to some quality among the multitudes of features that make up the elusive essence of John Paul II. Writing "religion" next to "pope" thus would not demand our casting our essentialist net particularly wide, but writing "war" certainly would. In this sense the more unexpected the pairing is, the broader the circle of definitions it requires from us, and thus the more original and perceptive, on some level, it may seem.

Moreover, it means that many pairings are, in principle, interchangeable. Schwarzenegger got stuck with war and Tom Cruise with money, but Schwarzenegger could have gotten "money" and Cruise "war." Even Susan Sontag ("thought!") could have gotten either "money" or "war." Sure, that would have prompted us to jump up and to argue, but, even as we argued, our essentializing brain would have been hard at work expanding its circles of meaning for "Sontag" and "money" or "Sontag" and "war" and groping for the contexts in which those meanings would intersect. Hence the paradox underlying Mr. Black's idiosyncratic referencing system. The action that seems utterly reductive—that is, narrowing the person to just one word—can actually be an exercise in conceptual amplification. "Reduction: expansion!"

PART TWO

Why Robots Go Astray, or The Cognitive Foundations of the Frankenstein Complex

1. What Is the Frankenstein Complex?

Isaac Asimov defines the Frankenstein complex as "mankind's . . . gut fears that any artificial man they created would turn upon its creator."[1] Although the particular character who offers up this definition in one of Asimov's short stories proclaims himself free of this complex and not afraid "that robots may replace human beings," at the end of the day he is proven rather spectacularly wrong. Indeed, many of Asimov's tales feature some variation of the Frankenstein complex being vindicated. His robots routinely disobey their makers, surprise them by doing things that they were emphatically *not* created to do, and even plot to destroy their masters. In fact, the whole genre of science fiction as we know it today is unimaginable without the recurrent motif of robots, cyborgs, animatrons, androids, virtual agents, and other artificial creatures going astray in a wide variety of shocking ways.

Let us widen the circle yet further. It seems that the Frankenstein complex was around even before the rebellion of the first official "Robot," of Karel Čapek's play *R.U.R. (Rossum's Universal Robots)* (1921), and before the murderous rampage of the "Creature" from Mary Shelley's *Frankenstein* (1816). Milton's *Paradise Lost* (1667) comes to mind, with its humans and angels defying their "Glorious Maker."[2] And then we are back to the Book of Genesis, in which it is God who has the Frankenstein complex. Upon learning that Adam and Eve have disobeyed him and eaten from the forbidden tree, God reasons that the "man has now become like one of us, knowing good and evil" and that he "must not be allowed to reach out his hand and take also from the tree of life and eat, and live forever." God is just as fearful of competition and displacement by the creatures that He himself has created as are Asimov's brilliant engineers at the fictional corporation "U.S. Robots." The deviant fictional robots are typically recycled, and, likewise, God banishes Adam and Eve and makes sure that they will ultimately turn into the dust of which Adam was originally made.

The recurrence of the "robot-gone-astray" motif suggests that we find perennially interesting stories about creatures who disobey their makers. We tell them to score a specific political point; to issue a vivid ecological warning; or to put a popular actress on the screen in the role of a sexy cyborg. Whatever our initial motivation and notwithstanding the rich array of subsequent interpretations we serve up, one thing remains the same: we tell such stories and apparently cannot get enough of them.

Of course, there are plenty of explanations as to why we find them so gripping. Such explanations typically invite us to see *Frankenstein, R.U.R.,* and *I, Robot* (1950), *Blade Runner* (1982), *He, She and It* (1991), and *Idoru* (1996) as responding both to fundamental psychological needs—such as the "obsession" with our "own image" or the fascination with "our own capacity for destruction"—and to specific cultural preoccupations of their time.[3] Thus Harold B. Segel considers Čapek's *R.U.R.* a characteristically expressionist reaction to technological expansion in the wake of World War I. In this view, Čapek's dystopian drama "embraces the common Expressionist vision of humans so overwhelmed by machine culture that they become subservient to and are ultimately swept away by the very robots they have created to lighten their own burden."[4] Similarly, J. P. Telotte suggests that the sci-fi movies of the seventies, such as *THX 1138* (1971), *The Clones* (1973), *The Terminal Man* (1974), *Westworld* (1973), and *Futureworld* (1976), hit a cultural nerve by contrasting the "mechanically crafted body powdered by artificial intelligence" with the "real, the genuine, the human." Telotte believes that such a contrast was particularly relevant in the era "that saw the creation of the first functional synthetic genes," the "rapid spread of reproductive technologies like the Xerox machine," and the "introduction of the videocassette recorder"—technological developments that blurred the "distinction between the originals and their copies."[5] And Asimov himself thinks that with the widespread use of the microchip in the eighties, a "brand new variety of technophobia" uncoiled itself in evil-computer stories, in which the increasingly "compact, versatile, complex, capable," and "intelligent" machines threatened to "replace not just a person but all humanity."[6]

These are good explanations, but for me they are not good enough anymore. Here is why. First, I am put off by the ease with which we come up with them. Indeed, it seems that *any* historical epoch or cultural niche can supply a whole slew of reasons why at this particular moment people would be particularly interested in the stories of artificial creatures who disobey their masters. You tell me that people loved the rebellious robot stories in the America of the 1970s because such stories both expressed and assuaged their anxiety about the blurred "distinction between the originals and their copies." I will happily counter that people also happened to love such stories in Soviet Russia in the 1970s (e.g., *Prickliuchenia Electronika* [1979]) because . . . let me think—ah!—because they allowed them to express their unhappiness about being treated by the state as easily replaceable parts of a huge mechanism. I will then quote you a series of everyday Soviet proverbs to the effect that nobody is unique ("nesamenimych u nas net!") and that we all are but nails, and bolts, and screws, easily replicated and easily replaced. Such wise sayings must have scarred several generations of people growing up under Soviet rule, and as such they must have intensified their hankering after fictional stories in which replicable robots turn out to be unique, irreplaceable, special, *human*.

Don't you find this theory compelling? I find it pat and irresponsible. It took me three seconds to come up with it, and you will never be able to either prove it or refute it. What kind of "explanation" is it?

My other quarrel with such explanations is that they leave out completely the human brain-mind that evolved to deal with natural kinds and artifacts but not with artifacts that look and act like natural kinds.[7] I wonder, then, what evolved cognitive adaptations we call on to conceptualize these strange creatures, or, to put it another, better, way, what features of such hybrids may reflect the workings of our evolved cognitive categorization processes?

Keeping these two objections in mind, I would thus like to advance the three following hypotheses. First, the fictional stories of artificially made creatures tease in particularly felicitous ways our evolved cognitive adaptations for categorization, and as such they remain perennially interesting to human beings. I use this "panspecies" language on purpose to challenge you to think of any human culture that does not have fictional stories about living creatures who are not born but "made," artifact-like.

Second, certain plotlines in such stories, for example, the "rebellious robot" plotline, could be seen as expressions of our intuitive attempts to resolve the state of cognitive ambiguity that has been forced on us by the challenge of processing the representations of such hybrid creatures. This is an important point, and in what follows, I will spend quite a bit of time on it. Here, however, is something that I want you to do in advance of that longer explanation. The next time you watch a science fiction movie or read a science fiction story, I want you to pay attention to that movie/story's use of *functionalist* language. I predict that science fiction narratives that develop the theme of a "rebellious creature" will foreground functionalist rhetoric, that is rhetoric emphasizing that a given protagonist—a robot, a cyborg, an android—was put together artifact-like, that is, it was *made,* or *custom-tailored,* or *designed* with a *specific function* in mind. I suggest that by cultivating such functionalist rhetoric to describe an entity that we at the same time cannot help but perceive as a living being, such stories tacitly render the motif of its subsequent rebellion particularly *cognitively plausible* or *cognitively satisfying.* This is not to say that our fictional man-made living beings *always* have to rebel: for every stray android of *Blade Runner* there is a perfectly-content-with-its-lot 3-CPO from *Star Wars.* But if they do rebel, we have been prepared for it by having struggled tacitly with their counterintuitive ontological status since the beginning of the story.

My third hypothesis concerns the current practices of historicizing our fascination with rebellious artifacts. I believe that the concept of the human brain becomes meaningless once you attempt to separate it from the culture in which that brain develops (and by the same token the concept of human culture becomes meaningless once you try to extract the human brain from it). As Frank Keil puts it, once we "acknowledge real structure in the causal and relational regularities of living beings, a structure that is largely distinct from other sorts of natural and artificial kinds, [we can] further see how cultural practices and beliefs interact with those structural patterns, and how each of those in turn interact with cognitive biases and constraints."[8] So fictional stories of rebellious robots indeed touch—sometimes intentionally, sometimes not—certain cultural nerves (e.g., an anxiety about encroaching technology, an unhappiness of being thought of as replaceable and replicable), but the very reason that they can touch these nerves is that they have first "grabbed" us cognitively, that is, they have created a zone of cognitive uncertainty and possibility.

To illustrate this latter point, let us first borrow an argument from part 1. I suggested in my discussion of *Speak, Memory* that had its narrator claimed outright that as an adult he finally understood that the essential feature of the Mademoiselle was her helpless kindness, it might have provoked an ambivalent reaction in readers. We might have felt that this is not quite it and that Mademoiselle was more complex than this, especially after everything we learned about her through the eyes of Nabokov's younger self. This is to say that even though we intuitively ascribe essences to a given individual (for doing so allows us to make inferences on the basis of minimal data), we remain particularly open to suggestion that we have missed grasping her essential feature once more. This may sound paradoxical, but it makes sense once you remind yourself that essences are bound to remain elusive because our quest for them reflects the quirks of our cognitive architecture and not the objective presence of essences in people.

This instability implicit in our essentializing may explain, at least in part, why a fictional story that features a person whose "self" is defined exclusively through his social position (e.g., "he is a lackey—that's all there is to him"[9]) implicitly prepares us for later complications: a revelation, for example, that the "lackey" actually has ambivalent feelings that transcend that initial characterization. Now, it does not mean that this later complication of the plot *has to* happen, merely that we have been *primed* for it (in case the writer does decide to develop it) by a strong statement of what the character "essentially" is.[10]

Let me restate this cognitive claim in historical terms: a fictional story that features a person whose "self" is defined exclusively through his social position lends itself to all kinds of subversive cultural readings grounded in specific ideological concerns of the moment. As soon as the story claims that social class defines the person, we are primed to look for culturally specific contexts that would subvert that claim.

On this level of analysis, it becomes very difficult to sustain the separation between the "cultural" and the "cognitive." To quote Keil again, "Conceptual domains are embedded in cultures that have strong influences on their natures, boundaries, and level of differentiation."[11] We have to speak about the ways in which the cognitive reinforces the cultural and the cultural reinforces the cognitive. This inquiry into the mutual cognitive-cultural reinforcement is quite different from explanations that focus on one (arbitrarily chosen) historical circumstance and ignore cognition altogether. (I gave you an example of one such explanation when I came up with a facile theory that my peers and I loved the movie about a rebellious boy-robot, *Prikliuchenia Electronika,* because even at our tender age we already felt the crushing power of the Soviet state.)

But before we approach specific stories of rebellious robots from a cognitive perspective, I need to elaborate some points of evolutionary research introduced in part 1. Let us turn once more to the work of cognitive psychologists and anthropologists.

2. *On Zygoons, Thricklers, and Kerpas*

The best thing about figuring out a set of rules is that it allows us to make a better sense of the cases that defy those rules. If we accept that our cognitive evolutionary heritage prods us to think of living kinds in terms of their essences and of artifacts in terms of their functions, we have a new way of approaching numerous cultural representations that violate the boundaries between living kinds and artifacts and thus challenge our primary ontological categories. In the already-discussed *The Essential Child,* Susan A. Gelman looks briefly at such boundary-violating entities, from crying statues and works of art to children's toys and cartoons. Two other recent books, Pascal Boyer's *Religion Explained: The Evolutionary Origins of Religious Thought* (2001) and Scott Atran's *In Gods We Trust: The Evolutionary Landscape of Religion* (2002), focus on the role played by such conceptual hybrids—for example, artifacts whose special magical powers imply sentience—in the world's religions. In what follows, I build on Gelman's, Boyer's, and Atran's studies to explore the categorical violations that abound in our fictional narratives. To do so, however, I first need to spell out the implicit *inferences* that structure those violations and make them so eminently *tellable:* fascinating, memorable, and always open to new interpretations.

To clarify the concept of inference, consider the two following sentences offered by Boyer: 1) "Zygoons are the only predators of hyenas," and 2) "Thricklers are expensive, but cabinetmakers need them to work wood."[12] If I ask you what will happen to a zygoon if I cut it in two, you will tell me . . . —now take a second to consider your answer before you read the end of this sentence and then compare your answer to mine—that a zygoon will probably die. If, however, I ask you what will happen if I cut a thrickler in two, you may speculate that it will be damaged or, on the contrary, will double its value because a thrifty carpenter can now use up only one half of a thrickler to make a cabinet and save the other half for the next job. Whatever you say about the halved thrickler, you will *not* assert that it will die. That fate will be reserved for a zygoon. I may ask you other questions—for example, whether zygoons can be made of rubber, or whether thricklers propagate by laying their eggs in the warm sand—and I won't even bother to write out here what I think your answers will be because I am *that* certain that I can guess them correctly, even though this is the first time that you and I have heard about zygoons and thricklers (and it's likely to be the last, too).

Now think of "kerpa." Let's say I tell you that to make a certain dish, you have to take one pound of boiled spinach, two teaspoons of butter, a pinch of salt, and a spoon and a half of kerpa. If I then ask you whether kerpas tend to smile sadly when you ignore their polite inquiries, or whether they can be folded down at the press of a button, you will say that they don't and they can't (unless, that is, we are in the world of Lewis Carroll, a subject of part 3). On the other hand, if I ask you whether you can take some kerpa on a plane in a plastic jar, you may respond with a tentative yes, a more definite answer hinging on finding out how perishable or how legal a substance kerpa is.

Are you impressed by your ability to converse with confidence about these previously unencountered entities: zygoons, thricklers, and kerpas? How do you explain this ability? To answer this question, Boyer reminds us that

> Minds that acquire knowledge are not empty containers into which experience and teaching pour predigested information. A mind needs and generally has some way of organizing information to make sense of what is observed and learned. This allows the mind to go beyond the information given, or in the jargon, to produce *inferences* on the basis of information given.[13]

So when we hear that zygoons are the only predators of hyenas, we immediately—though by no means consciously—place this new entity known as zygoon into the ANIMAL section of our mental encyclopedia, ANIMAL being a subcategory of a larger ontological category of living beings. Our essentializing mechanism then kicks in, enabling us to *infer* that "if you cut a zygoon in two it will probably die" (*for it is in the nature of most animals to die when they are cut in two*); that "zygoons cannot be made, they are born of other zygoons"; that "zygoons need to feed in order to survive"; and so forth.[14]

Similarly, when we learn that thricklers are expensive, but cabinetmakers need them to work wood, we assume that thricklers are TOOLS, a subcategory of artifacts. We think of artifacts as "manmade, inanimate, and defined by their function."[15] Moreover, we know that other, familiar items in this category, such as telephones, "cannot grow or eat or sleep" or lay eggs in the hot sand, "nor can screwdrivers or motorbikes." We will thus assume that "thricklers cannot grow or eat or sleep" or lay eggs in the hot sand.[16]

Finally, when we are told that a spoon and a half of kerpa is required for a recipe, we classify kerpa as SUBSTANCE. Substances can be weighed and carried around—as water, sand, and rocks—but they certainly cannot be folded down at the touch of a button (that quality is reserved for artifacts) and they cannot smile sadly when you ignore their polite inquiries. Only people can do that.

3. Theory of Mind

This latter point is so important and will come up so frequently in this book that it is worth our while to interrupt our present discussion to elaborate it. When we hear that a certain entity is capable of a mental state, such as feeling disappointed and trying to cover it with a sad smile (a smile that might also be implicitly signaling that feeling of disappointment), we assume that we are dealing with a sentient being, quite likely a human being. For—and here we have yet another instance of essentialist thinking—it is *in the nature of human beings to exhibit such ambivalent states of mind.* (I am not saying, by the way, that other animals cannot feel ambivalent, in a broad sense of the word, but rather that no other animal is capable of consciously *performing* its ambivalence for the sake of the spectator.)[17]

The mental condition of wishing both to convey and to cover one's disappointment is just one of the myriad mental stances available to us that we can recognize in ourselves and in others. Cognitive scientists have a special term, or actually several interchangeable terms, to describe the cognitive adaptation that allows us to consider a broad variety of mental states—thoughts, beliefs, desires, intentions—as underlying our own and other people's behavior. Atran and Boyer sometimes refer to it as folk psychology; other scholars have called it theory of mind or mind-reading ability.

The ability to interpret people's behavior (here, a smile) in terms of underlying mental states (here, a rather complex gamut of emotions) seems such an integral part of what we are as human beings that one may wonder why we need to dignify it with not one but three fancy terms (folk psychology, theory of mind, and mind reading) and elevate it into a separate object of study. The reason that over the last twenty years theory of mind has received the sustained attention of cognitive psychologists is that they came across people whose ability to "see bodies as animated by minds" was drastically impaired—people with autism.[18] By studying autism and a related constellation of cognitive deficits (such as Asperger syndrome), cognitive scientists and philosophers of mind began to appreciate our mind-reading ability as a special cognitive endowment, structuring our everyday communication and cultural representations. On the whole, studies in autism suggest that we do not just "learn" how to communicate with people and read their states of mind. People with autism, after all, generally have as many opportunities to "learn" these things as you and I. Instead it seems that we also have an evolved cognitive architecture that makes this particular kind of learning possible, and if this architecture is damaged, a wealth of experience would never fully make up for the damage.

What are the criteria used by psychologists to decide whether a given subject has an impaired theory of mind? In 1978, Daniel Dennett suggested that one effective way to test for the presence of normally developing theory of mind is to see if the child can understand that someone else might hold a false belief—that is, a belief about the world that the child knows is manifestly untrue. The first false-belief test was designed in 1983 and has since been replicated many times by scientists around the world. In one of the more widespread versions of the test, children see that "Sally" puts a marble in one place and then exits the room. In her absence, "Anne" comes in, puts the marble in a different place, and leaves. Children are then asked, "Where will Sally look for her marble when she returns?" After four—the age that apparently constitutes a crucial threshold in cognitive development—the vast majority of normal children pass the test, responding that Sally will look for the marble in the original place, thus showing their understanding that someone might hold a false belief.[19] By contrast, only a small minority of the children with autism do so, indicating instead where the marble really is. According to Simon Baron-Cohen, the author of the groundbreaking study *Mindblindness: An Essay on Autism and a Theory of Mind* (1995), the results of the test support the notion that "in autism the mental state of belief is poorly understood."[20]

Cognitive evolutionary psychologists working with theory of mind think that this adaptation must have developed during the "massive neurocognitive evolution" that took place during the Pleistocene (1.8 million to 10,000 years ago). The emergence of a theory of mind "module" was evolution's answer to the "staggeringly complex" challenge faced by our ancestors, who needed to make sense of the behavior of other people in their group, which could include up to two hundred individuals.[21] Baron-Cohen points out that "attributing mental states to a complex system (such as a human being) is by far the easiest way of understanding it," that is, of "coming up with an explanation of the complex system's behavior and predicting what it will do next."[22] In other words, mind reading is both predicated on the intensely social nature of our species *and* makes this intense social nature possible. (Lest this argument strikes you as circular, think of our legs: their shape is both predicated upon the evolution of our species' locomotion *and* makes our present locomotion possible.)

We engage in mind reading when we ascribe to a person a certain mental state on the basis of her observable action (e.g., we see her reaching for a glass of water and assume that she is thirsty); when we interpret our own feelings based on our proprioceptive awareness (e.g., our heart skips a beat when a certain person enters the room, and we realize that we might have been attracted to him or her all along); when we intuit a complex state of mind based on a limited verbal description (e.g., a friend tells us that she feels sad and happy at the same time, and we believe that we know what she means); when we compose an essay, a lecture, a movie, a song, a novel, or an instruction for an electrical appliance and try to imagine how this or that segment of our target audience will respond to it; when we negotiate a multilayered social situation (e.g., a friend tells us in front of his boss that he would love to work on the new project, but we have our own reasons for believing that he is lying and try to turn the conversation so that the boss, who, we think, may suspect that he is lying, will not make him work on that project and yet will not think that he didn't really want to, and so forth).[23]

Attributing states of mind is the default way by which we construct and navigate our social environment. This does not mean that our actual interpretations of other people's mental states are always correct—far from it! For example, the person who reached for the glass of water might not have been thirsty at all, but rather might have wanted us to think that she was thirsty, so that she could later excuse herself and go out of the room, presumably to get more water, but really to make the phone call that she didn't want us to know of. Still interpreting someone's mental state incorrectly is very different from not being able to conceive that there is a mental state behind the observable behavior. The former happens to all of us, the latter only to people with neurological disorders within the autism spectrum.

We read minds constantly whether we are aware or it or not. This is a crucial point, which can be easily obscured by our imperfect terminology. The words "theory" in theory of mind and "reading" in mind reading are potentially misleading because they imply that we go around thoughtfully communing with ourselves along the lines of, "Aha, I see that she is smiling, so I can infer that she must be pleased about something or wants somebody to think that she is pleased" or "She is reaching for a glass of water. That means she might be thirsty, or she may want me to think that she is thirsty."[24] Obviously, we do *not* talk to ourselves or others like that, except in certain specific situations, when for some reason we need to make more explicit our implicit perceptions, for example, "Well, I *assumed* that she was thirsty because she was glancing at that empty pitcher all the time, which is why I got up to fill it up again. Who knew that she was looking at it because she was thinking of hurling it at me?!"

In fact, it might be difficult for us to appreciate at this point just how much mind reading takes place on a level inaccessible to our consciousness. For it seems that while our perceptual systems "eagerly" register the information about people's bodies and their facial expressions, they do not necessarily make all of that information available to us for our conscious interpretation. Think of the intriguing functioning of the so-called mirror neurons. Studies of imitation in monkeys and humans have discovered a "neural mirror system that demonstrates an internal correlation between the representations of perceptual and motor functionalities."[25] What this means is that "an action is understood when its observation causes the motor system of the observer to 'resonate.'" So when you observe someone else grasping a cup, the "same population of neurons that control the execution of grasping movements becomes active in [your own] motor areas."[26] At least on some level, your brain does not seem to distinguish between you doing something and a person that you observe doing it.

The system of mirror neurons, which is responsible for activating the parts of your brain that would have been activated had you performed the action yourself, may "underlie cognitive functions that are as wide-ranging as language understanding and Theory of Mind."[27] After all, theory of mind does not simply kick in fully formed by the magical age of four: it develops gradually.[28] Thus it is possible that newborn infants' ability to imitate facial expressions of their caregivers—an ability apparently underwritten by mirror neurons—constitutes an early stage of the maturation of theory of mind.

In other words, our neural circuits are powerfully attuned to the presence, behavior, and emotional display of other members of our species. This attunement begins early (since some form of it is already present in newborn infants) and takes numerous nuanced forms as we grow into our environment. We are intensely aware of the body language and facial expressions of other people, even if the full extent and significance of such awareness escape us.

As a complement to research on mirror neurons, consider, too, the work on "interactional synchrony" by William Condon and his colleagues, who filmed a variety of communicative interactions and then analyzed frame by frame the body motions and speech sounds of participants. As William Benzon reports in *Beethoven's Anvil: Music in Mind and Culture* (2001), Condon discovered not only what he calls "self synchrony," that is, "the relationship between a person's speech patterns and their body movements: head, shoulders, arm and hand gestures, and so on," but also "interactional synchrony," that is, the "relationship between the listener's body and the speaker's voice."[29]

Whereas the existence of self synchrony is not particularly surprising ("after all, the same nervous system is doing both the speaking and the gesturing, and the cortical structures for speech and manipulation are close to one another"), the emergent synchrony between "the listener's gestures [and] the speaker's vocal patterns" is a rather striking phenomenon. It turns out that the "listener's body movements lag behind the speech patterns by 42 milliseconds or less (roughly one frame of film at 24 frames per second)," and that "infants exhibit near-adult competence at interactional synchrony within 20 minutes of birth."[30] All this shows once more that our minds and bodies are attuned to the minds and bodies of people around us to a much higher degree than we are consciously aware of.

Benzon further observes that

> Condon and others have also investigated interactional synchrony in children suffering from various pathologies, including dyslexia and autism. . . . They have observed dyslexic children whose right side would [respond] within the normal 42-millisecond period, while the left side would [respond] with the same sound at a delay of 100 to 266 milliseconds. Autistic children were similar, except that it is the right side that is delayed. . . . The ability to match one's movements to another's [thus] seems to be a condition of normal interaction with others. When this capacity is hampered, as it is in dyslexia and autism, communication is compromised. Synchrony creates a space of communicative interaction, a coupling between two brains in which they can affect one another's internal states. Interactional synchrony is not conscious or deliberate; it is not something one thinks about. It just happens, at least for most of us.[31]

As I sit in the library and work on my introduction to a volume on eighteenth-century British philanthropy, struggling with words and thus intensely focused on my own thought processes, a strange girl passes by. Something in me registers the direction of her walk ("She must want to go to the second floor"), her relaxed posture ("She is not in a hurry"), an umbrella under her arm ("She must have thought it would be raining today"), and the cut of her shirt ("She must have wanted to show off those well-toned arms"). And this is only what I am consciously aware of—God knows what else in her posture and facial expression my neural circuits are picking up on even though I do not care about this girl, most likely will never see her again, and won't recognize her if I do see her. We read minds nonstop, however biased, unintentionally wrong, intentionally muted, and just plain useless our attributions of states of mind to ourselves and to others might be.[32]

You see now that when I refer to theory of mind and mind reading in this book, I touch just the tip of an iceberg. That is, I speak about the aspect of mind reading that we can register and articulate—if "kerpa" smiles sadly, there *must* be a certain mental state behind that facial expression, which in turn means (and here our essentialist biases kick in promptly) that an entity known as a "kerpa" must be a sentient being. Just *how* we intuit the presence of a state of mind behind those distended facial muscles and what complex neural circuitry underlies our awareness of the kerpa's body language is a large issue, research-in-progress for years to come. What is important for the purpose of my argument is that attributing mental states to other people—as one expression of our profound neural-social awareness of them—is not something that we have conscious control of and can turn on and off as our fancy strikes us.[33]

4. Theory of Mind and Categorization: Preliminary Implications

Some preliminary implications of the research on theory of mind for our study of categorization become clear right away as we review the differences in our cognitive processing of living kinds (including persons, animals, and plants), artifacts, and substances. Let us say that we take a certain entity X to be a living being, a person. This categorization opens up possibilities for "more specific processing over the folk-psychological domain."[34] That is, we can now engage in mind reading and scrutinize X's behavior as indicative of certain mental states. We may decide, for example, that a strange smile playing on her face conveys a feeling other than straightforward satisfaction; or we may interpret her body language, when she throws herself on the sofa the minute she enters the house, as both indicative of her exhaustion after a long walk and of her intention to show us just how profoundly tired and thus incapable of doing any further chores she is at this point.

If we categorize a given entity as an animal (e.g., a cat), we can do some *limited* processing over the folk-psychological domain. We can infer, for example, that when the cat hisses at her owner and tries to bite and scratch him, she is angry at him for leaving her alone for several days.[35] The majority of our inferences, however, will belong to the folk-biological domain, that is, we will think of this specific cat in terms of what is "natural" to her species. We may keep in mind, for example, that cats need a certain kind of nourishment for survival; that they prey upon some animals and feel endangered by others; that they have certain reproductive and recreational needs; and then adjust our behavior toward this specific cat accordingly.

If we categorize a given entity as a plant (e.g., cabbage), we will do further processing exclusively over the folk-biological domain, for cabbages, too, have features "natural" for their species. For example, if we want to cultivate a patch of cabbages under our window, we will keep in mind the conditions under which cabbages, as a species, tend to thrive and adjust our watering and fertilizing strategies accordingly.

If we categorize a given object as an artifact (e.g., a bicycle), it enables us to engage in further processing over two conceptual domains.[36] The first domain is that of folk psychology, but with a very limited application focusing on the intended function of the object. We assume that the bicycle was made to fulfill a certain function: that is, it was its maker's intention that it would serve as a means of transportation. The second domain is that of folk mechanics: we proceed to figure out that the bicycle cannot stand on its own unless propped up against something; that it is separate from the ground on which it lies; that it weighs more than a telephone but less than a telephone pole to which it is currently chained and so forth. In contrast to our reasoning about persons, if we want to explain to somebody why our bicycle is lying on the ground, we will not ordinarily say that it must have lain down because it was exhausted after a long ride and wanted to show us just how exhausted it was; we may observe instead that it must have fallen down because it was not propped properly against the wall or that the object to which it was chained weighed less than the bicycle itself.

Finally, if we categorize a certain entity as a substance (e.g., quartz), we open up possibilities for further processing exclusively over the domain of folk mechanics. For example, we figure out what happens to quartz if we place it in water (will it slowly dissolve? will it sink?). As a point of contrast to our conceptualization of artifacts, thinking of quartz as found in the state of nature does *not* automatically activate any aspects of our folk-psychological domain, for quartz does not have a function and as such does not convey any information about the intention of somebody who "made" it. If it does, that is, if we start using this particular piece of quartz as a paperweight, we reclassify it as an artifact, thus adding to our folk-mechanical thinking about quartz some elements of folk-psychological thinking about its function.[37]

How far back into childhood can we trace this processing of perceptual data by different domains?[38] Here is a study featuring five-month-old infants that highlights the early age at which the differentiation between living beings (here, specifically humans) and artifacts manifests itself. Recent research by Valerie Kuhlmeier, Paul Bloom, and Karen Wynn suggests that in five-month-olds, this differentiation may be so strong that it fosters certain erroneous perceptions about the material nature of people. Kuhlmeier and her colleagues modified the well-known earlier experiments that show that infants expect objects to follow the principle of continuous motion. As evolutionary psychologist Jesse M. Bering puts it, "like any material substance, human bodies cannot go from A → C without first passing along the trajectory of B. For inanimate objects, infants are surprised (i.e., look longer) when the object disappears from behind one barrier and then seems to reemerge from behind another nonadjacent barrier. In the case of a human who violates the law of continuous motion, however, 5-month-olds are not surprised."[39]

It appears then that "infants are mistaken about the physical constraints that apply to humans," that is, they "do not readily view people as material objects." The existence of the "human/inanimate distinction, and the differential application of principles to each, may [thus] help infants to define these areas of knowledge early in development." Moreover, "the appreciation that these construals overlap—that in certain regards, people *are* just objects—may be a developmental accomplishment."[40]

Still, even after that "developmental accomplishment" has occurred and we have indeed come to understand that "in certain regards, people are just objects," this mature new understanding goes only so deep. For *even then* we remain fascinated by cultural representations that pointedly portray people as objects. Such representations remain interesting for us precisely because they activate inferences, from different conceptual domains, that cannot be reconciled with each other.

5. *Concepts That Resist Categorization*

Observe the self-enforcing circularity of our inferential processes. Imagine having just begun reading a fictional story and learning that its protagonist, whom we know only as "Andrew," forgot certain numbers. Even more specifically, we learn that "it had been a long time," but "if he had wanted to remember, he could not forget," but "he had not wanted to remember." Upon hearing this, we infer that Andrew is a person, for it is *in the nature of people only* to be capable of such complex mental states as wanting to forget something. Categorizing Andrew as a person immediately opens up the possibility that he has plenty of other, so far undiscussed, and most of them never to be discussed, thoughts, desires, and attitudes, for, again, it is in the nature of people—even if they are fictional characters—to have them.[41] For example, we are thus prepared to hear that Andrew used to feel particularly close to his maternal grandmother; that he hates SUVs; that he suspects that his second wife thinks that he is somewhat emotionally unavailable; that he has a good sense of humor; and so forth. This kind of information would conform to and reinforce our earlier assumption that Andrew is a human being, and that, in its turn, would make more information of this kind conceptually welcome.

And so it will go on and on in an agreeable feedback loop—as it does in countless works of fiction—unless we happen to read a work of *science* fiction, in which case we may suddenly learn that the number that Andrew has forgotten is his own serial number. And since it is certainly not in the nature of people to have serial numbers—that particular feature being reserved for artifacts—our hitherto smoothly operating feedback loop comes to a screeching halt.

Let us embroil ourselves in the same conceptual conundrum starting from a different point. Imagine having just begun reading a fictional story and learning that a certain entity, known to us as "NDR," had been "manufactured" a long time ago. So far so good: the word "manufactured" immediately leads us to classify this entity with artifacts and assume that NDR is man-made, inanimate, and has a specific function. Whatever that function is, we are prepared to hear about it, for it will agree with what we have already inferred about NDR. We are riding our feedback loop and are ready to be reinforced in our perception of NDR as an artifact, which would in turn prepare us for learning about other artifactual qualities of NDR.

But something unexpected takes place. We are told that NDR had a serial number—which is still perfectly fine and acceptable—and then, that he *forgot that number,* in fact, "had not wanted to remember" it. This puts an insuperable block in the path of our hitherto smooth inferential process. No matter how much extra information we may be given now that *confirms* our inferences about our NDR as an artifact—we could be told, for example, that it was "smoothly designed and functional"—the bit about his wanting to forget something sticks out like a sore thumb because it activates a whole new series of inferences that we associate not with artifacts but with living beings, and, most immediately, with human beings.

It does not matter, in other words, what our starting point is—we may begin by assuming, based on the existing textual evidence, that a certain entity is a human being, only to discover to our dismay that this human being has a serial number; or we may begin by assuming that a certain entity is an artifact, only to be confronted soon thereafter with this artifact's capability of wanting to forget things—in either case, we arrive at the juncture at which we have to readjust our feedback loop quite drastically.

How do we do it? Can we do it at all? Not really, according to Atran and Boyer, who argue that events and entities that violate our intuitive ontological expectations are *never* fully assimilated by any one ontological category. As such, they retain our interest, and stay in our memory, and remain perennially open to new interpretations (as do supernatural agents and magical artifacts in religions across the world). As Atran observes,

> The assignment to one of the primary ontological domains fails because further processing in accordance with intuitively innate expectations about folkmechanics, folkbiology, or folkpsychology is blocked. For example, the Arab talisman, European black magic crystal ball, and Maya sastun are naturally inert substances or artifacts with supernatural animate and sentient powers. As such, they violate intuitive ex-

> pectations about folkbiology and folkpsychology, and cannot be assigned to PERSON or ANIMAL, not can they remain simply SUBSTANCE or ARTIFACT. Talkative, vengeful, or pensive animals cannot be assigned to either PERSON or ANIMAL; omniscient but bodiless gods cannot be assigned to PERSON.

Moreover,

> By violating innate ontological commitments—for example, endowing spirits with movement and feelings but no body—processing can never be brought to factual closure, and indeterminately many interpretations can be generated to deal with indefinitely many newly arising situations. Notice that bringing processing to factual closure does not require actual verification, but only the reasonable possibility of such verification. Someone who is heard but not seen is visible in principle, whereas an invisible being is not.[42]

Boyer proposes the term "counterontological" to describe the entities that resist ontological closures. His reason for introducing this neologism is that the more familiar term "counterintuitive" can be easily read as "strange, funny, exceptional, or extraordinary," whereas the entities in question are not even necessarily surprising. For example, if my culture has a concept of a man who did not die after he was crucified, I "cannot really register puzzlement or astonishment every single time" that man's fate is mentioned. "It becomes part of [my] familiar world" that there are persons around who do not die when killed. But this concept is still counterontological because it includes "information contradicting some information provided by ontological categories," that is, that living beings, and people in particular, die when they are killed and do not come back.[43]

By failing to be assimilated by the category that it has initially activated—in this particular case, a living being, a person—a counterontological entity remains a promising source of new interpretations or, to make the same point somewhat differently, a fruitful source of new stories. Moreover, to return to the title of my book, these new stories will be profoundly structured by the ontological violation in question.

This point is crucial, so the next section will consider it in some detail. First, however, we need to address a potential misunderstanding that may arise when we juxtapose the notion of the counterontological with the assumption that ours is an essentializing species in a world without essences.

Counterontology implies ontology. That is, the notion of violation implies that there is a certain rule that can be violated. So when we think of the fictional robot as having complex mental states, we must also begin to essentialize that robot because we habitually essentialize beings who can have mental states. And in essentializing this robot, we go further than we usually do when we essentialize some artifacts in our everyday life (e.g., when we feel that something about their previous owner has "rubbed off" on them). But—and herein lies the potential for misunderstanding—how can one violate any rule by essentializing *anything* (in this case, the robot) if there are really no essences in the world? In other words, how can we call an entity counterontological if the particular ontology that it presumably violates is false (or nonexistent) to begin with?

You can see how, put this way, this issue may become a problem that will dog us until the end of this study. It needs to be resolved, and I propose resolving it by establishing an affinity between what Boyer calls ontologies and what Gelman calls heuristics. Remember again that Boyer deals not with entities out there existing independently from our perception, but with *cognitive* ontologies—that is, the ways of processing information about the world grounded in the particularities of our cognitive architecture. Similarly, Gelman observes that the belief in essentialism "can be considered an unarticulated heuristic rather than a detailed, well-worked-out theory."[44] But underarticulated, or even plain wrong, essentialism still gets us through the day—not always in the best shape, but what else is there for us to fall back onto?

This is to say that to the extent to which essentialism is the only game in town, it can be considered a cognitive ontology. And so from now on, when I talk about this particular violation of ontological beliefs, I mean the violation of our underarticulated, hazy, plain wrong, often harmful, but nevertheless cognitively enduring and *to that degree* ontological beliefs.

6. . . . and the Stories They Make Possible

Each culture has infinite forking subsets of possibilities—we can call them situation-specific cultural scenarios—associated with the event of death. For example, if a healthy man in his thirties, who was alive yesterday, is found dead today, we would expect a different behavior on the part of his family and his community based on the circumstances of his death. If he was a soldier and was killed in the line of duty, it is likely that his family will grieve for him. That grief may be accompanied by anger, if his family members think that the death was preventable (as in the case of friendly fire) or that the cause of the war was unworthy; by pride, if they think that his brave service to a worthy cause has benefited the whole community; or by relief, if when he was alive and among them, he treated them cruelly. If the deceased was a civilian, and there are reasons to suspect foul play, the grief may be accompanied by a publicly expressed desire for revenge, which should eventually be carried out by designated communal institutions. If the person died as a result of political persecution, say, in a concentration camp sponsored by the ruthless party in power, the grief may be accompanied by fear for other members of the family, whereas the desire for revenge may be tempered by the realization that the community at large should not suspect that the relatives of the dead man dare to harbor such subversive feelings.

The nuances of each situation modify the emotional responses of the people involved. Because the nuances are infinite, so are the responses; but *infinite* in the case of emotional reactions does not mean *arbitrary*. All possible emotional scenarios are grounded in the never-explicitly-articulated but nevertheless unquestionable ontological assumption that for our species death is final, and dead people do not come back to life.

Think then what happens to all of the above scenarios if the man in question certainly died, was dead for three days, and then came back from the dead. Will his family members still feel grief mingled with anger, or desire for revenge, or fear? Or will they experience half grief-half joy? Or mostly joy? Or joy mixed with desire for revenge? Or joy mixed with some fear? Will they try to conceal his return and pretend to be grieving and clamoring for revenge? And what will be the meaning of revenge? Would the murderer be sentenced to die, or would he be sentenced to die in the same strange sort of way, so that he too could come back in three days? Or would he be publicly commended for committing the act of murder and thus making it possible for his victim to come back in such a glorious fashion? And can we really call the murdered man a victim in this case? And is he really a man, given that he managed what no other human being ever did?

Violations of ontological expectations thus seem to be ripe with narrative possibilities. Turn to any realm of ordinary human experience (social, emotional, ethical), and consider it in the light of such a violation—and there is a story waiting for you. Of course, not every violation would do. As Boyer point outs, some of them are more inference-rich than others, "which makes the difference between good and bad stories."[45] Imagine, for example, that the same day returns again and again—which would be a violation of our intuitive expectations about the "nature" of time—then flesh out that counterontological idea by trying to envision a predicament of an ordinary man stuck in that day.[46] You may end up with an exciting movie, such as *Groundhog Day* (1993). On the other hand, imagine that the same day returns again and again, and then try to envision a predicament of a particular magical being, for whom time flows backward six days a week and sideways on Tuesdays, stuck in that day. I am not saying that it is absolutely impossible to muster enthusiasm for such a story featuring an "overdose" of ontological violations (working with cognitive approaches to fiction teaches one pretty quickly not to opine on what is narratively "impossible"), but it is rather difficult. Piled on thickly, ontological violations are subject to the law of diminishing returns; their narrative potential may slide downward quickly.[47]

To clarify further the issue of inference-rich violations of our intuitive ontological expectations, consider the relative narrative value of the three following stories:

> 1. Once upon a time there was a darning needle. One day it was used to mend leather slippers, and it broke. The owner of the slippers dipped the broken needle in hot sealing wax and stuck it to her blouse as a brooch. Soon thereafter, however, it fell out of the blouse and down into the sink. It landed at the bottom of the gutter, and it still lies there.

I think you'll agree with me that this story is not terribly exciting. The possibility of your remembering it for its own lively sake and presenting it to others as an entertaining little tidbit is slight. We can envision situations in which you will repeat it to somebody to make some larger point. For example, if you are a thrifty person wishing to warn a friend who uses darning needles for wrong purposes, you may offer this story up as a cautionary tale. Or, if we imagine a community where needles are scarce and function as status symbols we can use that story to illustrate an adage along the lines of sic transit gloria mundi.[48]

You can see what contortions I am going through to endow the chronicle of the darning needle with some narrative appeal. The truth is that every turn of this story merely confirms our intuitive expectations about artifacts; for example: artifacts have functions; at times they change their functions; if dropped, a tiny object is likely to get lost; artifacts are subjects to laws of gravity. As such, this little vignette calls for an absolute minimum of new interpretations and leaves us with no burning unanswered questions, except, perhaps, "Who cares?" and "What's the point?"

But let us change that story slightly:

> 2. Once upon a time there was a darning needle. One day it was used to mend leather slippers, and it broke. The owner of the slippers dipped the broken needle in hot sealing wax and stuck it to her blouse as a brooch. Soon thereafter, however, it fell out of the blouse and flew up in the air. It still floats over the kitchen table.

This narrative is distinctly more interesting because it violates our intuitive expectations over the domain of folk mechanics. By defying gravity, the needle emerges as a strange little object whose behavior prompts numerous questions, for example, will the thread float in the air too? Will the slippers? Was it something about the wax that did it? How will people in the house react to this event? What kind of house is this? What is going on? The author's work is cut out for her now: she has to construct a narrative framework within which this violation of our intuitive ontologies makes sense. We shall feel distinctly dissatisfied if the story simply ends here.

Note that once this story is indeed extended to make sense of the floating needle, it is likely that we won't hear much about that particular artifact anymore. The gently hovering needle is not that fascinating by itself (Kubrick's *2001: A Space Odyssey* [1968] is not, after all, a chronicle of the hovering steel slab, however counterontological an object such a slab is). Its attraction lies in its being a part of something else, some possible world similar to and yet different from ours. It is the account of that world that we are hoping to get, or, to be more specific, it is the description of that world's inhabitants, their relationships, thoughts, and feelings that we are hankering after. Intensely social species that ours is, we always prick our ears at the possibility of hearing a story that would provide our theory of mind with relevant material to process.[49] The image of a floating needle seems to promise such a story.

Finally, consider the opening of "The Darning Needle"—a short story by Hans Christian Andersen—the way it actually appeared in print. The plot is exactly the same as in example 1, but the ontological violations chosen by the author open that plot up in all kinds of unforeseen directions:

> 3. Once upon a time there was a darning needle who was so refined that she was convinced she was a sewing needle. "Be careful! Watch what you are holding!" it shouted to the fingers who had picked her up. "I am so fine and thin that if I fall on the floor you will never be able to find me again."[50]

We still have some discourse here that agrees comfortably with our intuitive expectations about artifacts: the needle has a function; it is subject to the laws of gravity; its size makes it difficult to find it once it is lost. At the same time, our protagonist is a talking needle! A needle aspiring to a higher social status and wishing to convince others around her that she deserves to be treated as if she has already attained that exalted status! A self-deluded needle! Such a creature clearly cannot be assimilated by either the "artifact" or the "person" category. It will remain one of a kind—the needle with a distinctly human mind—and as such it immediately opens up a broad array of narrative possibilities.

What are those possibilities?

7. The Stories That Can Be Told about a Talking Needle

There were no voluble needles in the ancestral environment in which our tendency to essentialize living beings took shape. This means that once we hear that a given entity has a capacity for talking and complex mental stances, we cannot help but to start perceiving that entity in terms of what is "in the nature" of human beings. The story that opens with an image of a loudly ambitious needle can thus develop in infinitely many directions, all reflecting our intuitive ontological expectations about people.

People weave complicated webs of social relationships. They have friends, enemies, colleagues, associates, mentors, pupils, neighbors, servants, and masters. They receive inheritance, serve on jury duty, aspire to higher social status, get demoted, lie, get caught, reinvent themselves, fall in love, tell stories about themselves and others, dedicate their lives to important causes, waste their time, get married, entertain themselves and their friends, bore everybody around them to death, and so forth. Our talking needle can do all these things too.

Except that, just as in the earlier case of the man who comes back from the dead (a violation of our intuitive ontological expectations, which forces us to reconsider all familiar social scenarios concerning death in which such a man would figure), the details of the social life of a protagonist who is also an artifact would be adjusted to reflect that protagonist's ambiguous ontological status. Here are some possibilities. First, the needle's sworn enemy would have the capacity to stymie its love life by hiding it behind a stack of dishes on a kitchen shelf. Second, its career advancement ripe with delightful new social opportunities could be attained through having being broken in two and dipped in hot sealing wax. Third, its admiring retinue may include a thread trailing behind it. Fourth, its service on a jury may prompt other members of the jury to be particularly careful so as not to sit down accidentally on their sharp tiny associate.

I am inventing some of these scenarios and lifting others from the tales of Andersen—that doyen of fictions that explore conceptual hybrids such as ambitious needles, hopeful street lamps, and mumbling cucumbers. I want to emphasize again that as immensely playful, and creative, and unpredictable as these stories are, their unpredictability does not imply arbitrariness. It is true that their range of plots and subplots is infinite—as is the range of social situations open to human beings—but the actual narrative expressions of all those plots will be structurally constrained by the nature of ontological violations present in them. This is to say that an author can write the talking needle into practically any recognizably human social situation, but then he will have to adjust some aspects of that situation to reflect the unique nature of the protagonist. The resulting narrative will thus contain a number of thought experiments exploring the implications of situating a human mind in a completely inhuman body—a body, moreover, that has certain properties defined by the artifact's erstwhile function.

Stories of animated artifacts with complex and distinctly human social lives have a long pedigree. (After all, the readers of antiquity shared our cognitive biases and were as fascinated by images of conceptual hybrids as we are today.) One of my favorite examples comes from ancient Syria. On his fantastic travels, the narrator of Lucian's *True Story* (c. 166 A.D.) visits "Lamptown," a city that is located "in the air between the Pleiades and the Hyades, though much lower than the Zodiak." The inhabitants of Lamptown behave mostly like ordinary human beings, although they still die like . . . well, lamps:

> On landing, we did not find any men at all, but a lot of lamps running about and loitering in the public square and at the harbour. Some of them were small and poor, so to speak: a few, being great and powerful, were very splendid and conspicuous. Each of them had his own house, or sconce, they have names like men, and we heard them talking. They offered us no harm, but invited us to be their guests. . . . They have a public building in the centre of the city, where their magistrate sits all night and calls each of them by name, and whoever does not answer is sentenced to death for deserting. They are executed by being put out. We were at court, saw what went on, and heard the lamps defend themselves and tell why they came late.[51]

The social and emotional life of lamps continues to attract writers. Here is the synopsis of Pixar's short animation feature *Luxo Jr.*, created by John Lasseter in 1986:

> A baby lamp finds a ball to play with, and it's all fun and games until the ball bursts. Just when the elder Luxo thinks his kid will settle down for a bit, Luxo Jr. finds another ball—ten times larger. Luxo Jr. has a great dad in the larger lamp. Even though he is a bit unpredictable, the elder Luxo gives him room to grow and explore. And the tiny lamp has no problem with that.[52]

Read side by side, the 166 A.D. and 1986 A.D. "lamp" stories vividly illustrate my larger point about the simultaneously cultural *and* cognitive construction of our fictional representations. On one hand, the image of a socially engaged lamp titillates us and yet makes sense today for the same reason that it titillated readers and yet made sense to them nineteen hundred years ago: it forces us to perceive this artifact as belonging to the domain of human beings, thus activating our inferences about what human beings do and feel. On the other hand, certain crucial details of the stories mark them as deeply grounded in their respective cultural milieus.

Some of these details are emotional. The relationship between the protagonists of Pixar animation—"the elder Luxo gives [Luxo Jr.] room to grow and explore"—are described in terms intelligible and attractive to an audience belonging to a very specific place and time. Thus, I know that fifteen years ago, as a new immigrant from Russia, I would not have felt the appeal of this sentiment as strongly as I feel it today, now that I have learned to appreciate the value that my friends and colleagues put on the view that children should be given "room to grow and explore." It's not that this sentiment was inconceivable in Russia when I was growing up, but, among the people whom I knew, it was not held at the same premium as it is (and has been for some time) among many people in this country. And whereas I have no way of telling if Lucian's first readers strongly agreed with this sentiment, I suspect that even in his time it might have been more intelligible to readers of some social classes, economic means, and personal temperaments than to others.

Yet other details are technological. Lucian's readers could understand why putting the lamp out meant killing it. Today's viewers would hardly perceive switching Luxo Jr. off as killing him; they could read it as making him take his nap so that he would wake up refreshed and ready for new exploits, however ardently he might protest against "going to bed." The omnipresence of electricity has taken some of the drama out of the extinguished light. Sure, Luxo Jr. can still "die" but it would take more than simply switching him off or unplugging him (perhaps to kill the lamp today, you have to smash it or cut the cord?).

And so forth. At every turn, the tone of the "Lamptown" passage bespeaks a sensibility that differs from the sensibility animating the story of Luxo Jr. and his father. Yet the fascination with the image of an animated artifact and the patterns of such an animation remain the same (i.e., people "loiter" and play, and so does an animated lamp; people have families, and so do animated lamps; people die, and so can an animated lamp) because they are rooted in the biases of our cognitive information processing that has remained stable over these two thousand years.

A True Story is considered one of the earliest surviving examples of science fiction. Let us now turn to more familiar examples of this genre and apply to them what we have learned about our cognition of categorization. Using our new terminology, I suggest that the protagonists of these stories—robots, cyborgs, androids—are *counterontological* entities who straddle the respective domains of artifacts and living beings. I further argue that at least one recurrent plot turn of our science fiction stories—the plot of a rebellious robot—can be traced to our drive to restore the broken conceptual feedback loop and to fit an apparently counterontological entity within a comfortably familiar ontological category.

8. Asimov's "The Bicentennial Man"

Remember Andrew, who "had not wanted to remember" his serial number? Asimov's short story "The Bicentennial Man" (1975) begins to build on—and tease—our intuitive ontological assumptions about living beings and artifacts from page one. We meet Andrew Martin when, apparently sad and, in fact, "driven to the last resort," he visits a doctor's office. This information provides enough input for our essentializing proclivity to get in gear and enable us to categorize Andrew as a human being.

"Facing" Andrew "from behind the desk" is a "surgeon." This also gives us enough information to conceptualize this other character as a human being. The nameplate on the surgeon's desk includes a "fully identifying series of letters and numbers, which Andrew [does] not bother with. To call him Doctor would be quite enough."[53] A slightly strange bit, this one, about the series of letters and numbers identifying a person, but we do not let it override our initial assumption that the surgeon is a human being. As a default assumption (for doctors who face us across their desks in our world are still human, even with the advance of new technologies), it needs stronger evidence to be overruled.

Moreover, we are riding a comfortable feedback loop with Andrew. His decision not to "bother" with the surgeon's identification tacitly reinforces our initial perception of him as a person, and so does his next action:

> "When can the operation be carried through, Doctor?" he asked. The surgeon said softly, with that certain inalienable note of respect [yes, certainly a person] that a robot [WHAT?!] always used to a human being [right, Andrew *is* a human being]: "I am not certain, sir, that I understand how or upon whom such an operation could be performed."[54]

The revelation that the surgeon is a robot is somewhat shocking (though now we can account for that peculiar bit about "letters and numbers" on his nameplate), and Asimov rushes on to strengthen our perception of the surgeon as an artifact:

> Andrew Martin studied the robot's right hand, his cutting hand, as it lay on the desk in utter tranquility. The fingers were long and shaped into artistically metallic looping curves so graceful and appropriate that one could imagine a scalpel fitting them and becoming temporarily, one piece with them.
>
> There would be no hesitation in his work, no stumbling, no quivering, no mistakes. That came with specialization so fiercely desired by humanity that few robots were, any longer, independently-brained. A surgeon, of course, would have to be. And this one, though brained, was so limited in his capacity that he did not recognize Andrew—had probably never heard of him.[55]

Note the rhetoric of functionalism seeping through the description of the robot. Its fingers are "shaped" in a certain way to enable it to excel at its function, which implies a "shaper"—a "maker" behind this particular artifact. The word "brained" intensifies this impression, for we do not literally "brain"—"give brains" to human beings (we do brainwash them, but this is a metaphorical use of the functionalist language, to be discussed separately).

As the conversation continues, Asimov finds yet another way to stoke our functionalist thinking about the surgeon:

> Andrew said, "Have you ever thought you would like to be a man?"
>
> The surgeon hesitated a moment as though the question fitted nowhere in his allotted positronic pathways. "But I am a robot, sir."
>
> "Would it be better to be a man?"
>
> "It would be better, sir, to be a better surgeon. I could not be so if I were a man, but only if I were a more advanced robot. I would be pleased to be a more advanced robot."[56]

"So as to be able to better perform his function," we infer. Like any artifact, the surgeon *is* what he does, and his ambitions are limited to doing that one thing well. Note here—lest we forget why this tracing of rhetoric of functionalism is important—that this rhetoric in all its different guises would not have been effective in amplifying our impression of the surgeon as an artifact had it not been for our evolved cognitive tendency to think of artifacts in terms of their function. This rhetoric makes sense—as does a story built around that rhetoric—because it appeals to our evolved cognitive heritage.

"Obedience is my pleasure," drones the surgeon; and it seems that by the end of this introductory scene Asimov almost succeeds in making us reconceptualize as an artifact the character that we initially perceived as a living being. I say "almost" because as a talking and "independently-brained" machine, this robot still activates in us certain inferences associated with living beings and so remains a somewhat counterontological entity.

The issue of gradience is important here. At the end of their conversation, Andrew Martin makes a startling revelation that he, "too," is a robot. Still, it is clear to us that Andrew's ontological status is more ambiguous than that of the machine he is talking to. Andrew comes across as "more" counterontological than the surgeon: he seems to be capable of a broader range of emotions, and we have not yet heard anything about *his* function (though as an artifact he is bound to have one). If I were to ask you now, jumping ahead of my argument, which of the two robots is more likely to "rebel" against his masters, you would certainly point to Andrew. It is not that a rebellion is inconceivable in the case of the surgeon, but it would require some catastrophic event (e.g., a short circuit) to wake him up from his happy artifactdom.

Having first established Andrew as a human being only to then drop the bombshell of his "too" being a robot, Asimov proceeds, in the second part of the story, to build up our impression of his protagonist's ontological ambiguity. He does this by mixing rhetoric that activates inferences belonging to the domain of artifacts with rhetoric that feeds into our essentialist biases. We hear, for example, that when first "manufactured," Andrew was "smoothly designed and functional," with a "serial number NDR—," only to learn immediately after that he forgot and "had not wanted to remember" his serial number. An artifact who wants to forget that he is an artifact (which implies a fully developed theory of mind) cannot be assimilated by our cognitive systems along the same lines as a chair, a toaster, a microchip, or even a fellow machine completely satisfied with its present status (i.e., the surgeon).

The information that is supposed to signal most strongly Andrew's status as an artifact—the reference to his function—is couched in purposefully ambiguous terms. We are told that Andrew "had been intended to perform the duties of a valet, a butler, a lady's maid." The pointedly asexual ring of this job description (valets and butlers tend to be men whereas lady's maids, women) reminds us how meaningless gender differences are for a machine. But, by the same token, the *triple* function that forces one to move from tasks associated with one gender to tasks associated with another seems to mark Andrew as a priori more versatile than, say, a robot surgeon who is made to excel at just *one* task. This tentative impression of Andrew's versatility is strengthened in the next paragraph, in which he is portrayed as happily ignoring his original function, for the children of his owners "would rather play with him" than use his services as a valet.[57]

Andrew's "rebellion," which unfolds gradually and is nonviolent, consists of transcending his original design and becoming as unpredictable and multifunctional as (if not more so) any human being. First, he turns out to be a brilliant artist. His woodwork earns him a small fortune and later enables him to "buy his freedom" from his owners, even though he continues to live close to them and be treated almost like a family member. Increasingly independent as Andrew is, however, his robot status still makes him vulnerable, for any human stranger can order him around and force him to do things harmful to himself. To address this incongruence, Andrew's friends lobby for a law that would protect him, and, eventually, a series of laws are enacted "which set up conditions, upon which robot-harming orders were forbidden."[58]

Meanwhile, Andrew undergoes a series of surgeries (many of them made possible by his innovative research in prostheology—his new area of interest), which turn him into a more "organic" creature—one who looks, eats, and breathes like a human being. His work with "prosthetized devices" takes him to the Moon, where he is "in charge of a research team of twenty human scientists," while robots treat him "with the robotic obsequiousness due to a man."[59] Finally, Andrew insists on one more operation, this time to render himself irrevocably mortal: a suicide of sorts, but, nevertheless, the step that convinces the "World President" that Andrew should be granted his lifelong ambition and be officially declared a human being.[60] The story ends as Andrew lies dying; wishing that his last thought would be that he "was a man" but thinking instead of the late daughter of his original master, whom he adored—a romantic failure to control his emotions, which should conclusively mark him as a person and not a robot.

Andrew's actions frighten and antagonize his makers, the powerful corporation U. S. Robots. Especially in the early days of his quest for humanity, the corporation tries to lay its hands on Andrew so as to dismantle and study him. When that fails, they revamp their entire enterprise to ensure that the new machines that they produce will never be as flexible and troublesome as this renegade robot. For, nonviolent as Andrew is, his behavior seems to vindicate the Frankenstein complex. Andrew grows smarter and richer than the human beings who "manufactured" him, and he learns to deploy a wide range of emotional and legal tactics to make them do what he wants. At one point he is aware of having contributed to the "lying" to and "blackmail" and "badgering and humiliation" of the head of U. S. Robots; at another, he finds himself setting "stern conditions" to that man's "stunned" successor.[61] (In both cases, Andrew's goal is to move closer to becoming human.)

I see Andrew's rebellion as "cognitively satisfying" for us, the readers, because its outcome—the erstwhile robot's official admission to the ranks of humanity—eases the cognitive tension built up by the story. Andrew is a narrative construct that activates inferences from two different conceptual domains: the domain of living beings and the domain of artifacts. By making Andrew go through his life as a half-man and a half-machine—a hybrid state reflected in the fact that he has complex emotions but "man-made" features—Asimov forces his readers to straddle these domains.[62] By bringing his rebellion to fruition and showing Andrew die "as a man," he finally allows us to conceptualize him as a human being, or, if not really and truly human, then at least, as more human than before.

With this in mind, perhaps we can redefine from a cognitive perspective the psychological phenomenon known as the Frankenstein complex. When we encounter a fictional character whose ontology seems to pull us in two different directions, we intuitively grapple for the ways to restore at least one of our broken feedback loops (for we cannot restore both) and to resolve the cognitive ambiguity by conceptualizing that hybrid as either a living being or an artifact. But what kind of plot development would allow us to achieve this kind of cognitive resolution?

To become more "artifactual" in our eyes, the hybrid has to signal strongly his rigid functionalism. The surgeon in "The Bicentennial Man" does by showing his inability to think of himself outside his profession. To become more of a person, on the other hand, the hybrid has to emerge as humanly multifunctional, which means that he has to be perceived not in terms of one limited function but in terms of some ineffable, invisible, and ungraspable "essence" (as Andrew seems to be by the end of the story). To do so, however, he has to rebel against the inhumanly narrow role foisted onto him by his makers, a rebellion that could be violent or peaceful, depending on the context of the story. Our anticipation of this rebellion—which could otherwise be seen as an expression of our subconscious hope for a resolution of the conceptual problem that we have on our hands—is what we call the Frankenstein complex.

9. Cognitive Construction of "Undoubted Facts": "The Bicentennial Man" and the Logic of Essentialism

As we follow Andrew's metamorphosis from a machine to a "man" and note the rhetoric used to negotiate the difference between a robot and a living being, we may get an uncanny feeling that either Asimov was familiar with the research of Atran, Gelman, and their colleagues, or that those scientists read a lot of Asimov prior to embarking on their studies. As one example of the resonance between those studies and the logic of essentialism cultivated by "The Bicentennial Man," consider Gelman's report of a conversation that her colleague, Francisco Gil-White, had with a group of Kazax men in western Mongolia:

> In the midst of conversation about nature-nurture conflicts, Gil-White asked the following: "If I stayed here, and learned Kazax, and Kazax customs, married a Kazax girl, and became a Muslim, would I not be a Kazax?" The respondent's reply was: "Even if you do everything like a Kazax, and everybody says you are a Kazax, you still aren't a real Kazax because your parents are not Kazax. You are different inside." And he pointed to his chest.[63]

This conversation about what makes one a Kazax, with its strong underlying assumptions about the "hidden, nonobvious properties that impart identity," is both profoundly familiar and yet striking precisely because of its easy familiarity and its seeming "naturalness."[64] We see the man's point—simply *cannot help* seeing it because of our own essentialist proclivities—even if we do not particularly like that point or cannot explain what it is exactly about being born to Kazax parents that makes you a Kazax (there is no "gene" for "Kazaxness," after all, no more than there is a "gene" for being a Ukrainian or Latvian).

Look now at the scene in "The Bicentennial Man" in which, after returning from the Moon, Andrew approaches the head of the legal firm that represents him and asks whether it would be possible to be "legally identified" as a human being. As he puts it,

> On the Moon, I was in charge of a research team of twenty human scientists. I gave orders that no one questioned. The Lunar robots deferred to me as they would to a human being. . . . I have the shape of a human being and organs equivalent to those of a human being. My organs, in fact, are identical to some of those in a prosthetized human being. I have contributed artistically, literarily, and scientifically to human culture as much as any human being now alive. . . . Why, then, am I not a human being?

The lawyer replies that, "treated as a human being by both robots and human beings," Andrew is thus a "human being *de facto*" and he should be satisfied with that, since any attempt to establish him as a human being *de jure* "would run into human prejudice and into the undoubted fact that however much [he] might be like a human being, [he is] not a human being."[65]

To me, the most interesting aspect of this conversation is the lawyer's appeal to "the undoubted fact" of Andrew's essential nonhumanness. Generally, claiming that something is an "undoubted fact" may have the opposite effect of making the audience question the truth-value of the statement and suspect some interested motive in the speaker.[66] So when we have the lawyer portraying himself as the last word on the subject (which, the story implies, he is not), we are inclined to think that he is narrow-minded and prejudiced and perhaps wrong. We have been thus inched further along toward perceiving Andrew as human.

As far as essentializing goes, this is a curious moment. What the lawyer says is grounded in our essentialist biases and as such is something that many people would actually find quite commonsensical. After all, the world in which one can look, and act, and be treated as a certain kind of person and still not *be* that kind of person is the same old familiar world in which "even if you do everything like a Kazax, and everybody says you are a Kazax, you still aren't a real Kazax," unless you were born a Kazax. In both cases, our evolved cognitive tendency to essentialize living kinds naturalizes a distinction that otherwise would be difficult to defend either logically or scientifically (e.g., a Kazax versus a non-Kazax; an android who is thoroughly "like" a human—in fact, almost "more" than human—versus a "real" human). So, what Asimov does here is take an everyday essentialist notion and put it in the mouth of an arrogant lawyer—a somewhat suspect authority to begin with—to the effect of undermining the validity of this notion on the grounds of suspect authority.

But undermined does not mean eradicated. The essentialist reasoning about what Andrew "really" is or is not continues to inform the story, in fact, has to, for this is the kind of story that exploits our underarticulated but enduring essentialist intuitions from its first to its last line.

Hence, soon after the conversation with the lawyer, another, more appealing authority (an older congresswoman) attempts to clarify the reason for Andrew's difference. The clarification is pointedly antiscientific but it is made to sound plausible because of the way it feeds our essentialist biases. When Andrew persists in seeking some logical explanation of what it is that makes him so irrevocably different from human beings, he is finally told that the proof is in the brain: "Your brain is man-made, the human brain is not. Your brain is constructed, theirs developed. To any human being who is intent on keeping up the barrier between himself and a robot, those differences are a steel wall a mile high and a mile thick."[67]

Andrew still refuses to simply bow to the wisdom of this quintessentially essentialist view (which, incidentally, recruits the rhetoric of functionalism to signal

Andrew's artifactual nature) and wonders instead if one "could get at the source of [people's] antipathy" to the constructed brain. To this, his friendly interlocutor says "sadly" that it is "the robot in [him] that drives [him] in [the] direction [of] trying to reason out the human being."[68] Again, this is a thoroughly essentialist sentiment, for it relies on our acquiescing (without even noticing it) that it is "natural" for a robotic living being to be stubbornly logical in a situation in which a "real" human being would apparently just agree that humans are unreasonable and leave it at that.

Note that I am not saying this to condemn the cruel humans who would not let "poor Andrew" in the club of "true" humanity.[69] My point is that every turn of the story seems to make sense because of the unspoken essentialist assumptions that we bring to bear upon it. If by some impossible magic we were to remove that essentialist propping, the story would collapse as if robbed of its skeleton. (Of course, it would collapse equally fast, if not faster, were we to remove our unspoken assumption that there is a state of mind behind each action of a fictional character. Our cultural representations, including our fictional narratives, are structured by our intuitive ontological expectations on an untold variety of different and yet interlocking levels. In fact, it seems that the more we find out about the workings of our evolved cognitive architecture, the more daunting becomes the challenge of developing a conceptual framework that could grasp the overall pattern of interactions among these levels.)

Moreover, to come back to my facetious suspicion that either Asimov read too much Gelman or Gelman read too much Asimov, consider the similarity between the story's pontification about Andrew's brain as the seat of difference between him and "real" humans and our everyday reasoning about what constitutes the "essence" of a person, her "soul," her "core," as reported in *The Essential Child*.

It seems that, to support our essentialist thinking about individuals, we tend to look for the "hidden nonobvious causal properties that impart identity" in such bodily organs as the heart or the brain. So people who undergo heart-lung transplants sometimes comment on certain personality changes, suggesting that some of their donor's "essence" has transmigrated to them. A wife of a heart donor says about meeting the recipient of her dead husband's heart: "I could feel his essence, his energy." Her interviewer (not a cognitive psychologist) chimes in: "Anyone who receives a new heart is getting a big ball of subtle energy. Ancient cultures have known about subtle energy throughout history, and have viewed it as the vital force of all creation." What Gelman finds interesting about such accounts is "not the question of whether or not it is true or even plausible, but rather how [the interviewed people] construct a personal essence: as nonvisible, internal, persisting through massive changes, and having the capacity to influence outward behaviors and preferences. It is both material (located in the heart, a flesh-and-blood bodily organ) and immaterial (an 'energy' or 'soul')."[70]

As the seat of the person's "essence," the heart figures frequently in fictional stories featuring counterontological creatures such as Tin Woodman from Frank Baum's *The Wizard of Oz* (1900) or Sharikov from Mikhail Bulgakov's *Dog's Heart*. Although in "The Bicentennial Man" Asimov opts for the brain as the locus of difference between Andrew and a "real" human being, the logic is exactly the same. We are told explicitly that the "seat of Andrew's personality is his positronic brain and it is the one part that cannot be replaced without creating a new robot"; so it is not really surprising that toward the end of the story, Andrew decides that the only thing that he can do to bring down that "steel wall a mile high and a mile thick" that separates him from humanity is to change irrevocably the nature of his brain.[71] As he puts it,

> See here, if it is the brain that is at issue, isn't the greatest difference of all the matter of immortality? Who really cares what a brain looks like or is built of or how it was formed? What matters is that brain cells die, *must* die. Even if every other organ in the body is maintained or replaced, the brain cells, which cannot be maintained or replaced without changing and therefore killing the personality, must eventually die.
>
> My own positronic pathways have lasted nearly two centuries . . . and can last for centuries more. Isn't that the fundamental barrier? . . . Human beings . . . cannot tolerate an immortal human being. . . . And for that reason they won't make me a human being.
>
> I have removed that problem. Decades ago, my positronic brain was connected to organic nerves. Now, one last operation has arranged that connection in such a way that slowly—quite slowly—the potential is being drained from my pathways.

Andrew's plan appears to succeed spectacularly. His decision to sacrifice his life to be human catches "the imagination of the world," and, shortly before his death, he is officially declared a man.[72] Yet note how ironic both his plan and its success are from a cognitive perspective: although we perceive both natural kinds and specific individuals in terms of their essences, any actual attempt to locate the seat of that essence is bound to remain inconclusive at best and ridiculous at worst.[73]And so it should be, of course, given that our essentialist thinking reflects particularities of the cognitive makeup of our species rather than certain objective truths about the world.[74]

Which means that Andrew's quest after the ever-receding essential difference between himself and human beings is, objectively speaking, meaningless even though it feels cognitively meaningful. Andrew may accumulate one human trait after another—human appearance, human creativity, prosthetic organs widely used by humans—neither any one of these traits by itself nor their combination (or "cluster," to use Kripke's term) will ever capture the essence of being human, for there is both too much and too little to capture when one hunts for the embodiment of an essence.

But if Andrew's quest is bound to fail in objective terms, it may succeed rhetorically, which in this case is even better. And so to begin substituting rhetoric for reality, Andrew takes as his starting point one essentialist assumption that has been dogging him lately—"the essence of the human being is the human brain"—and counters it with another: "the essence of the human brain lies in its mortality."

Andrew's theory of what constitutes the essence of the human brain is as plausible and indefensible as any essentialist theory. However, precisely because it is both plausible and unfalsifiable, the introduction of a strong appeal to the emotions of the people whom he is to persuade—his willingness to back up an indefensible essentialist assertion with his life—earns him the coveted name of man. This does not mean that these people have managed to recategorize Andrew as a human being in their own minds (this is a really murky area), but they have clearly become doubtful enough about his ontological status to finally accede to his plea.

Let me use a stark example to illustrate my argument that an appeal to emotions is more effective if an issue at hand forces us to think in essentialist terms. If I show you that I am willing to die to prove my point that pigs fly, you may think that I am crazy or overly passionate about my idée fixe, but your opinion about pigs' levitating abilities will hardly change. But if I show you that I am willing to die to prove my point that to be a human being is "all about" being free; or loving one's family; or being mortal, you will certainly be impressed enough to think, if only for a short time, "what if she is right?" Because essentialist thinking is both a powerful mental habit and an ontological bottomless pit, we seem to be particularly vulnerable to a rhetoric that appeals to emotions to argue about essences.

(Note, incidentally, that an emotional dynamic is also at play in the earlier episode with the lawyer: because an unappealing character offers the "verdict" about Andrew's nonhumanness, we are less inclined to agree with the verdict. The issue itself remains irresolvable but our emotions can be manipulated into making us think that perhaps there is a resolution and that it is quite different from what this lawyer claims with such tactless confidence.)

And *of course* Asimov did not have to read the research of cognitive evolutionary scientists to figure out that Andrew might want to appropriate the essentialist thinking of humans in order to make his own plea for human status compelling. Writers work with perfect assurance our evolved cognitive tendencies because constructing their stories this way makes intuitive sense to them. I knew that you would say that kerpas do not smile sadly when you ignore their polite inquiries because I myself know that they cannot (that is, as far as I can "know" anything about something that does not exist). That knowledge is both innate—that is, *made possible* by our evolved cognitive architecture—and cultural, for had you and I been born into an alternative world in which substances did smile, it would have overridden our innate propensities.[75] Similarly, in writing stories that rely extensively on our essentialist biases, Asimov "knew" that they would make sense for us because they made intuitive sense to him. Cognitive psychology and anthropology thus help to bring into the open implicit assumptions about the world, which make the three-way communication between the writer, the reader, and the text possible.

10. *Made to Rebel*

I asked you earlier the next time you read a story or watch a movie featuring robots, cyborgs, androids, and so forth to pay attention to the presence or absence of the rhetoric of producing such beings as artifacts, that is, the rhetoric of *making* them with a specific *function* in mind. You understand now why I wanted you to note it. I suggest that such rhetoric would be strongly emphasized in the stories that portray artificially made creatures heading toward rebellion.

This emphasis is far from accidental if we consider that our evolved cognitive architecture prompts us to recognize and interpret environmental stimuli according to those stimuli's perceived properties. When the properties of a given object seem to satisfy the input conditions of the domain of artifacts (we learn that it was "made"), certain inferences associated with the domain of artifacts (e. g., "artifacts have functions") are activated and we are thus primed to think about the narrowly defined "function" of this object. If at the same time, this object is portrayed as having a human shape or human emotions, we are faced with a cognitive conundrum.[76]

For we have on our hands a character whose ontology seems to pull us in two different directions—an artifact with a definite function and a living being, essentialized, multifunctional, and largely unpredictable. The resulting cognitive ambiguity can be resolved in several ways. One of them, familiar to us since Genesis, is to show this ontologically ambiguous creature rebel against its original function in order to align itself more strongly with living beings. Adam was "made" by God to till the Garden, and Eve was similarly "made" by God to help Adam—their ultimate rebellion against God's decree was thus strongly implied the moment this rhetoric of "making" was used to describe human beings.

A classic example of a being that straddles two ontological domains and can never be completely assimilated by either is of course the "Creature" from Mary Shelley's *Frankenstein*. As Lawrence Lipking puts it, Shelley presents the reader with "genuine, insoluble problems," one of which concerns the status of its protagonist: "Is the Creature a natural man or an unnatural monster?"[77]

For, on one hand, the Creature exhibits a fully developed theory of mind, which forcefully encourages us to perceive him as human; hence critical interpretations that emphasize his vulnerability, sensitivity, loneliness, and artistic temperament. On the other hand, we cannot forget about the "infinite pains and care" with which he was "formed" by his "creator"; about Victor's consciously "select[ing] his features"; about the "work" of muscles and arteries that could still be seen beneath his "yellow skin." Hence it is easy for us to assimilate the latter-day cinematic image that emphasizes the patchwork surface of this artifact. As Denise Gigante observes, "The stitches we can only assume are holding him together (a visual image impressed upon us by screen versions of *Frankenstein*) expose the mechanics of his creation."[78]

Victor Frankenstein expects that the "new species" that will emerge from his "workshop" will "bless" him and experience a profound "gratitude" toward him: "its creator and source."[79] But things made in workshops do not bless their creators. They don't hate them, either: they simply cannot have states of mind. And when the narrative continues to insist that they do, it presents us with a cognitive problem—a continual challenge to our intuitive ontologies.

We can thus say that the rebellion of the Creature and other "artificial" living things is "cognitively satisfying," this satisfaction being proportionate to the intensity with which the functionalist rhetoric of "making" a sentient being is emphasized throughout. Fictional narratives tease our evolved capacities for categorization and then try to mitigate the ontological uncertainty that arises from that teasing, as they depict the hybrid creatures acting rather more unpredictably than their makers expected and thus becoming more human.

Here are a few other specimens from my collection of artificially made creatures whose rebellion is strongly foregrounded by functionalist rhetoric. My fa-

Figure 4. Scene from *Making Mr. Right* (1987). Ulysses (John Malkovich) is talking to his maker (also John Malkovich). "I didn't program you for love! You were made to explore space!" "Well, I don't care that much about space anymore."

vorite is the goofy android Ulysses, played by John Malkovich in Susan Seidelman's movie *Making Mr. Right* (1987). Ulysses is made by one Dr. Peters, a brilliant if incurably geeky scientist, and he has a specific, narrowly defined function. Ulysses was built to be sent into space, where he will work on his own for many years, a feat of solitude presumably unbearable for any regular human being. Instead, the android falls in love with a woman and begins pursuing her in lieu of his educational program. Dr. Peters is outraged: "I didn't program you for love! You were made to explore space!" to which the android replies, "Well, I don't care that much about space anymore" (fig. 4).

Without giving away the story's ending, let me just say that Ulysses settles decisively on the side of humans and comes to embody any woman's dream man—"Mr. Right." This outcome must feel particularly satisfying to the viewer jolted by such cognitively challenging phrases as "I didn't program you for love." This phrase implies a human being who understands language and emotions and can be bullied by his social superior and yet who was "programmed" machine-like, with a particular function in mind. Ulysses can't be both—something's got to give. And what better way to prove that this impossible counterontological creature can, after all, be assimilated within our familiar categories than to show him falling in love, disregarding his function, and knowing that he is disregarding it—becoming *human* in other words?

To return very briefly to the argument of part I, *Making Mr. Right* also titillates our essentialist biases by having Malkovich play both Mr. Peters and Ulysses. We are thus given a pleasurable opportunity to observe the "twins" talking to each other and to register the differences in their mannerisms and attitudes (which we construct as indicative of some deeper essential difference). The physical similarity between the two characters also adds an extra layer to our ongoing task of assimilating Ulysses within one of our familiar ontological categories. Not only does this man-made object look like a human being, but he also looks like a specific human being, from whom, yet, he is apparently "essentially" different because his outward behavior betrays a different set of values and attitudes. The movie thus intertwines the challenge to our essentialist biases with the challenge to our categorizing biases. It builds on both, it titillates both, it eventually resolves both by assuring us that the android is indeed extremely human—and therein lies the "essential" difference between the two men, for it is his maker who turns out to be more like a robot.

Ulysses' rebellion is relatively peaceful. True, he disappoints and displaces his maker, but there is no active violence involved. In contrast consider the four-wheeled protagonist of Asimov's short story "Sally," whose transcendence of her function leads her on a murderous rampage.

"Sally" features a fleet of retired cars that can run on their own and are apparently endowed with personalities and souls. The narrator, an elderly man who takes care of these cars, is especially fond of one of them, named Sally, who has a particularly sweet disposition. The conflict implicit in the concept of a car with a *soul* (which, as we remember, is frequently associated in the Western tradition with essence) is articulated by an evil trespasser, Mr. Gellhorn, who claims that "cars are *made* to be driven."[80] The emphasis on the cars' being made and having thus only one, strictly defined, function relegates them to the domain of artifacts, negates the possibility of their having any agency or soul, and thus renders perfectly legitimate Mr. Gellhorn's wish to recycle them to build newer models.

The cars themselves, however, think otherwise. The story ends with them ganging up on Mr. Gellhorn and killing him, giving a premonition to the narrator of the story that a day may come when the cars will want to get rid of all people, not just the evil ones who want to hurt them. Predictably, the narrator does not "get as much pleasure out of [his] cars as [he] used to. Lately, [he notices] that [he is] even beginning to avoid Sally."[81] The irresolvable ontological ambiguity implicit in the idea of a car with a soul—irresolvable because arguably grounded in the particularities of our cognitive makeup—makes Sally a ticking bomb and the park for the retired cars a breeding ground for the future rebellion.

Čapek's *R.U.R. (Rossum's Universal Robots)*, the play that introduced the word "robot" into our lexicon, depicts a community of "working machines" who discover that, to have a "soul," they have to get rid of their makers/masters. Functionalist rhetoric is rampant throughout the play, and predictably, at the end, when the robots defeat their creators, they turn into humans themselves.[82] Or so we are encouraged to think in the closing scene, when the male robot offers to sacrifice his life to save a female robot from being taken apart, and both of them exhibit a suite of pointedly human emotional reactions.

In Philip K. Dick's novel *Do Androids Dream of Electric Sheep?* (1968) people are invited to move to Mars with a promise that as an incentive each settler will be given a "loyal, trouble-free companion," a "*custom-tailored* humanoid robot—designed specifically for YOUR UNIQUE NEEDS, FOR YOU AND YOU ALONE—given to you on your arrival absolutely free."[83] Those of you who have seen the movie *Bladerunner* (which is loosely based on Dick's novel) know just how trouble-free those androids of the future turn out to be. As a character in another of Dick's novels, *The Man in the High Castle* (1962), puts it, "no man should be the instrument for another's needs."[84] Androids in *Do Androids Dream of Electric Sheep?* are apparently able to dream, and dreaming is strongly associated with having a conscience, agency, a personality—qualities that we do not associate with objects made as "instruments for another's needs." This clash of rhetoric, already present in the title of the novel, is repeated again and again in the story proper, implicitly preparing the reader for the appearance of androids who refuse to obey their makers and to serve the "specific unique needs" of their owners.

11. *Why Phyllis Is Still a Robot*

Thomas Berger's novel *Adventures of the Artificial Woman* (2004) works the familiar "rebellious robot" territory. Or so it wants us to think as it pounds us with its functionalist rhetoric. Consider the opening of the novel in which a man turns to technology to help him out with his love life:

> Never having found a real woman with whom he could sustain a more than temporary connection, Ellery Pierce, a technician at a firm that made animatronic creatures for movie studios and theme parks, decided to fabricate one from scratch.
>
> The artificial woman would naturally be able to perform every function, but sex was the least of what Pierce looked for in his made-to-order model. . . . [Having been convinced by his long experience that] even the most amiable [woman] would [eventually] turn sarcastic, make aspersions on his tastes, disrespect his judgments, and

> in general be an adversary instead of an ally, [he began to suspect that] the solution might well be not, so to speak, human but rather technological.[85]

Any potentially troubling ethical issues are seemingly taken care of by the logic of triumphant functionalism. Because, as Pierce puts it, he "*made*" his perfect woman—literally "built her from scratch"—"it would not be criminal for [him] to neglect, discard, or even destroy the creature he had made, in any of which circumstances he would be guilty only of wasting much of his life."[86]

As cognitive literary critics, we know now that something else is going in these opening pages. The references to "making," "fabricating," and "building from scratch" a living creature tacitly prepare us for the future rebellion of that creature, and we can even map out the general domain of that rebellion. If the function of this particular Galatea—here named Phyllis—is to remain amiable and never be sarcastic or disrespectful toward her Pygmalion, then her uprising will manifest itself in some form of acknowledged intellectual superiority over him.

And, sure enough, on the last pages of the novel, Phyllis waxes profoundly disrespectful toward her creator and is suitably punished for her unseemly behavior. In fact, we first encounter what could be perceived as Phyllis's rebellion against her maker much earlier in the story, but because Pierce himself does not perceive it as such, we end up not thinking about it in these terms either. Here is what happens. Shortly after being activated and installed in the capacity of Pierce's wife, Phyllis decides to leave him to try to make it in the world on her own. Pierce's reaction to her decision is appealingly enlightened. The "feeling of pride" overcomes "any resentment [he] might have known." He sees Phyllis's "declaration of independence . . . as his triumph, not only technological but also moral, for he [has] not opposed her departure. He [has] created her, but he did not pretend to own her."[87] This is an amicable parting. As such it draws rather more on our familiar scenario of a reluctant parent letting go of a teenage kid than that of a maker trampled over by the robot he put together.

When Pierce meets Phyllis again several years later, she has behind her a brilliant career in the movies while he has spent those years as a homeless man eating out of a trash bin and living in a crawl space under a viaduct. (Phyllis's departure has unexpectedly broken his heart, rendering him unfit for normal functioning.) The reunion is cordial enough: Pierce is thrilled to see the woman he has loved and missed all these years, and Phyllis is apparently technologically unable to entertain any but warm feelings toward the man who made her. With Pierce as her manager (though all the creative ideas come from her and not from him) Phyllis first becomes a successful talk-show hostess and then runs for president as a write-in candidate and wins. Her animatronic identity is kept secret, though her inhuman forbearance certainly serves her well when she is faced with the grueling electioneering schedule. Pierce remains her husband, best friend, and political advisor, who sees her through every step of her campaign.

That admirable teamwork comes to a screeching halt when Phyllis begins to consider her first steps as the U.S. president-elect. She intends to appoint the least politically correct but, from her perspective, the most effective people to key posts in her administration, and when the appalled Pierce disagrees with her plans, she simply fires him. Phyllis informs Pierce that she will now need a different kind of partner and regales him with the following assessment of his worth:

> "Ellery, little by little I've come to realize you have a second-rate sensibility. Oh, you know your way around technology, but that's about your limit. I can see why you never succeeded with human women. Inferior as they are, they could still easily get your number."
>
> He shrugged. "That's true enough."
>
> "You admit it? You see how weak you are? I'm afraid you're a born bootlicker, Ellery. You're not much of a man, even by low human standards."[88]

Pierce is understandably "stung" by Phyllis's words and pays her back by luring her close to him under the pretense of giving her a good-bye kiss and then killing her by pressing the tiny "fail-safe button" inside her ear. He is convinced that the political appointments she was about to make "gave [him] no choice" except to turn her off permanently. Allowing her "to lead this country in [her] current state would be ruinous" for the people's well-being, and he, Pierce, would "become the mad scientist of the old horror movies instead of what [he is], a romantic with stars in his eyes." The novel ends with Pierce standing next to Phyllis's lifeless body and thinking that perhaps he could rebuild her, "if he could work the bugs out of what was manifestly a winning design."[89]

What is striking about these concluding words of the novel is that they convince us, at least for a while. We do not immediately think of Pierce as a murderer—indeed why should we, given that his victim is not a real woman but merely a "winning design"? We have identified with Pierce for most of the story—for isn't he our "human" mainstay in the world of animatrons, pimps, politicians, and other suspect creatures?—and so we continue to buy into his version of events. And according to this version, Pierce is a tragic figure, a romantic hopelessly in love with a beautiful machine that has been friendly enough but remained constitutionally unable to truly reciprocate his tender feelings and that now has to be offed before it acts on its Frankenstein impulses and harms the whole country if not the whole world. As Pierce sees it, and would have us see it too, Phyllis's *real* rebellion has finally taken place, and it consisted in rejecting the voice of human reason (i.e., Pierce's) and insisting on a course of political action that might have made sense to a robot but would be dangerous for its flesh-and-blood electorate. This robot has to be destroyed, as have been other countless creatures of science fiction, hell-bent on harming humanity with their logical but deadly designs.

Adventures of the Artificial Woman is thus a comedy of manners / political satire that seems to play out a typical "rebellious robot" scenario. Yet something else is happening here. That last bit about stars in Pierce's eyes is worthy of Nabokov's Humbert Humbert, another murderer laboring hard to convince his readers that he is really a starry-eyed romantic and the only true sufferer of his narrative. What we have on our hands, in other words, is an "unreliable narrator" story, in the manner of Nabokov's *Lolita* and Kazuo Ishiguro's *Remains of the Day*.[90] As James Phelan puts it in his *Living to Tell About It* (2005), the presence of an unreliable narrator in a work of fiction means that we "must reject the narrator's words and reconstruct an alternative."[91] And, as it turns out, *Adventures* contains plenty of evidence for a very different interpretation of what is going on between Phyllis and Pierce. What prevents us from registering that evidence early on, however, is the novel's manipulation of our categorization biases. Let us recover that interpretation and then figure out why it is that we tend to miss it the first time around.

Let's begin with the issue of rebellion. I have discussed so far two occasions on which Phyllis may be said to "rebel"—one when she leaves her maker and the other when as a new U.S. president she is about to challenge internal politics-as-usual. One of Berger's many jokes, however, is that there is a third, real rebellion in the story, but neither Pierce himself, nor the reader, recognizes it as such when it takes place. Phyllis was "fabricated" to remain amicable and respectful toward her human partner, to remain an ally, to refrain from casting aspersions on his judgments. Such was her function, and it is this function that Phyllis dispenses with immediately upon coming into consciousness. She begins to treat her creator as an idiot almost from the first week of their life together, but because while doing it she sounds properly submissive, nobody notices it.

For Phyllis knows the importance of talking the talk even when she is not walking the walk. She learns it after facing a mortal danger. Already in the first chapter, Phyllis informs Pierce that she will not harm him even though she is physically stronger than he (which is true), to which he responds, "You're not stronger than me. I can literally take you apart any time I want." This is a real threat, and Phyllis recognizes it as such. She then appears to be "deliberating." "'You're right, Ellery,' she [says] finally, winking at him. 'I wasn't serious. I was lying.' 'No, you weren't, Phyl,' Pierce assures her, 'machines have no sense of irony and therefore never joke and never lie. You were simply trying to take power. Automobiles try that from time to time, with striking accelerators, brakes that fail, and so on.'"[92]

What is most amazing about this exchange is that we buy—at least for a while—Pierce's inane explanation of Phyllis's behavior. Instead of realizing that as a sophisticated self-educating system, she has already transcended her original mental stasis and is now manipulating Pierce in his self-satisfied blindness, we take it for granted that as a creator Pierce knows the limits of his creation. If he thinks that she is merely acting up like a temperamental automobile, then so it must be. Note, too, that Pierce is actually not the brilliant inventor familiar to us from other "Frankenstein" stories. He is a mere "technician at a firm that made animatronic creatures," which means that he has a rather limited knowledge of the creature he put together from the parts provided by his company. Of course, we ignore that important bit of information as we automatically agree with Pierce's description of Phyllis's mental processes.

Having missed the meaning of that early conversation, we continue to misjudge Phyllis's behavior and trust Pierce's assessment of what is going on. When shortly thereafter Phyllis decides to leave Pierce, we agree that it was both noble and open minded on his part to let her go. We do so in spite of the compelling evidence that by convincing himself that he is proud of Phyllis's "declaration of independence" and his own enlightened response to it, Pierce may simply be making virtue out of necessity. For to ensure her safe exit, Phyllis has first tied Pierce to the bed under the pretense of sexual foreplay, using "the belt, tie, and socks to fasten his wrists and ankles to the bedstead."[93] Pierce could not stop Phyllis even if he wanted to.

Moreover, it is clear that Pierce is only capable of admiring Phyllis's "spunk" while he perceives her not as an autonomous being but as an extension of himself. Because he has "made" her, her "great strengths" and the accomplishments that may come out of her strengths belong to him. This is a profoundly possessive stance, and we will see it at play throughout the novel. When the two meet after their long separation—Pierce fresh from the gutter, and Phyllis from movie stardom—the balance of power remains just as tipped as when he was lying supine tied to the bed and she was going out full of enterprising energy. Still,

Pierce persists in thinking of Phyllis as a mechanical extension of his genius and marveling at what he has wrought. And because the smart woman never says anything to upset his self-congratulatory beliefs—on the contrary, she keeps emphasizing that she is what Pierce "made" her to be—he never feels any jealousy or resentment at her mounting professional accomplishments.

And then that idyll ends—in decidedly human terms. Either carried away by her new political power or forgetting that Pierce still has the ability to destroy her (or not knowing in the first place how this ability may manifest itself), Phyllis finally stops massaging the fragile ego of her husband-creator. It might be that she is tired of a man whose powers of imagination continue to fall short of her own, but who, nevertheless, feels entitled to criticize her admittedly unorthodox yet compelling political ideas. Phyllis's dismissal of Pierce is couched in pointedly cruel and mocking terms (perhaps calculated to show him that she really means it), and it is also foolhardy: smart as she is, she should have anticipated his response. But then, the very foolhardiness of her behavior might be taken as evidence of her *human* frailty.[94]

Because Pierce, of course, would not be able to accept either her independence or her changed tone. His murderous response to Phyllis's dismissal of him thus emerges not as the heroic act of a savior of humanity, but as the revenge of a jealous man whose wife and social status are about to be taken away in one stroke. For Pierce knows that once separated from Phyllis he would no longer be able to claim the credit for what she does and assert that her accomplishments are really his.

Even more important, Phyllis now sounds similar to other women from Pierce's past, who, as we remember, tended to "turn sarcastic, make aspersions on his tastes, oppose his opinions, disregard his judgments, and in general be an adversary instead of an ally."[95] Pierce could not kill those bitchy females for putting him down, but he can certainly dispatch Phyllis since she is, as he never stops reminding us, a mere robot. Has he not stipulated at the onset of his narrative that "it would not be criminal for [him to] destroy" her, since in such "circumstances he would be guilty only of wasting much of his life?"[96] And—to take our "unreliable narrator" hypothesis to its logical end—isn't it possible that writing down the story after the fact (that is, after he has murdered Phyllis), Pierce, whether consciously or not, ratchets up the language of functionalism to absolve himself of his crime?

Thus if we succeed in reading the novel *against* Pierce's reading, he emerges as a possessive mediocrity and a murderer of a remarkable woman who was constantly thinking outside the box, who succeeded in every demanding career she tried, and whose unorthodox political vision just might have been beneficial for the country. With Phyllis dead, the post of the president will be assumed by her vice president, a nobody whose single life accomplishment—he once shot a robber in the eye—is very much to Pierce's liking. To this kind of uncomplicated toughness he can easily relate.

Now the interesting question is *how* Berger makes us believe Pierce's version of the story in spite of all the evidence against it. How is it that we see Pierce naked and tied down to his bed and still think that he can stop Phyllis from leaving but does not because he is so noble and open minded? Or, just as striking, how is it that we buy his idiotic insistence that Phyllis is not capable of irony, lying, or joking, when throughout the novel she lies, and jokes, and pretends whenever the occasion calls for it? (For crying out loud, the woman is an actress and an occasional movie director!—the professions that thrive on the ability to inhabit different states of mind and to pretend.) What is it that prevents us from recognizing Pierce as an unreliable narrator right away and reading skeptically everything he says about himself, about Phyllis's mental abilities, and about his own relationship with her?

Writers who portray unreliable narrators use a broad range of rhetorical techniques to pull wool over their readers' eyes. I have already discussed some cognitive dimensions of such techniques in my book on cognitive science and narrative, *Why We Read Fiction: Theory of Mind and the Novel* (2006). Here, I want to add to that earlier discussion by showing that Berger relies—intuitively but assuredly—on his readers' response to the rhetoric of functionalism to make them see Phyllis through Pierce's eyes: a robot, whose destruction matters only as far as Pierce has wasted some parts of his life on what turned out to be a bug-ridden machine.

We know that Phyllis transcended her initial function almost as soon as she was "born" and became a self-directing dynamic system, a quick learner and a creative thinker. As such, she is as "human" as she can get, especially considering that our definition of what it means to be human is hopelessly mired in our essentialism. For example, is a person with all artificial limbs still human? What if all of her internal organs are replaced with artificially grown ones? What if she is a clone? What if her DNA has been tampered with? What if her brain is hooked up to a computer? Trying to answer such questions and to explain why we answer them one way or the other demonstrates pretty quickly that our thinking on this subject has little to do with some "objective reality" of "true" humanness and everything with our essentialist biases. A clear-cut definition of what it means to be human in the age of ever-expanding prosthetic tampering is wishful thinking, an illusion.

But if Phyllis is human on so many levels, why then does it take us some time to register grief at her demise and to recognize Pierce for what he is—a criminal? The reason for that is that the shrilly insistent rhetoric of functionalism—all these references to Phyllis being "made," "manufactured," "built from scratch," and so forth—keep interfering with our assessment of her as a remarkably gifted person, and not just a sophisticated robot. The novel thus pushes its readers in two very different directions—forcing us to think of Phyllis as a human being and an artifact—and it never lets go of that conceptual conflict.

When I think of the effects that this conceptual conflict has on me as a reader, I realize, for example, that I consciously downplay Phyllis's unusual origins in order to be able to think of her as a brilliant woman murdered by a selfish non-entity. That is, I have to make an effort to ignore the rhetoric of functionalism tirelessly mouthed by Pierce because I know that this rhetoric affects me (e.g., it prevented me from immediately perceiving Phyllis's demise as murder). Similarly, whether you agree or disagree with my reading of Pierce as an unreliable narrator and Phyllis as a brilliant person, your interpretation will also be struggling with or giving in to the rhetoric of functionalism pervading the narrative. To fully appreciate this effect, let us compare *Adventures* with a novel that also deals with the relationship between a human man and an "artificial" woman but that significantly weakens the language of functionalism in describing that woman—William Gibson's *Idoru.*

12. . . . *and Why Rei Toei Is Not*

A lot is going on at the same time and in many different locations—geographical and virtual—in Gibson's novel *Idoru.* And so it should be because the author wants us to feel just as disoriented in the fast-paced universe of his story as many of his characters do. And the main reason that they do—even those who are usually at home in cyber- as well as any other space—is the rumor that the middle-aged but hugely popular rock star called Rez intends to marry a software construct from a Japanese virtual reality program. Because several cottage industries live off Rez's fame, their representatives would concern themselves as a matter of course with *any* woman who was about to enter his life on a long-term basis. This case, however, is quite special. Once it becomes clear Rez's fiancée is not likely to "polish his gun, pick up his socks, [and] have a baby or two," a great scramble ensues to find out what the hell is really going on (is Rez mad?), who is behind it (the Russian mafia?), and how it will affect everybody around him. And so one Colin Laney, a recently unemployed "serious netrunner," or "researcher," as he likes to think of himself, is hired by Rez's entourage and sent to Japan to find out who the lady is and if there actually *is* a lady.[97]

Slowly but surely Laney learns that the bride-to-be is an Idoru—an "idol-singer" named Rei Toei. She is a "personality-construct, a congeries of software agents, the creation of information-designers . . . akin to what . . . they call a 'synthespian,' in Hollywood." Plenty of people are eager to explain to Rez that Rei Toei does not really exist, to which he replies that he perfectly well knows that she doesn't and that they have no imagination. Meanwhile, Rez and the Idoru seem to be very much in love, and even Laney has to admit that Rei Toei, when she goes out as a "hologram . . . something generated, animated, projected," makes a rather striking dinner date and a nice conversationalist.[98]

Gibson meticulously calibrates his language of functionalism to describe Rei Toei. First of all, what is the function of a virtual singer? To sing? To entertain? To titillate the eye and the ear of young viewers hooked on the net? All this is already way too multifunctional and complex for your run-of-the-mill artifact. Second, the story of Rei Toei's origins—in stark contrast to Berger's "artificial woman"—eschews the crude vocabulary of "making," "fabricating," and "building from scratch." True, the Idoru is characterized as a "local product" and we occasionally hear of people who "worked on her design."[99] At the same time, when we get to meet one of those designers, one "Mr. Kuwayama" (who has to be present with some computer equipment whenever Rei Toei is out in the real world, in order to project her image), the functionalist rhetoric becomes slippery, laden with qualifications and ambiguities. Thus we learn that "Mr. Kuwayama is Rei Toei's creator, in a sense. He is the founder and chief executive officer of Famous Aspect, her corporate entity. He was the initiator of her project."[100]

Founder, chief executive, initiator of her project, creator *in a sense* . . . What happened to our plain old "maker" and "creator"?

And here is how "the initiator of her project" speaks about Rei Toei. No, excuse me, here is how he speaks about *something,* for he is giving an answer to a question that Laney (and thus the readers) happened to miss behind the din of the restaurant conversation. It is thus possible that he is not even speaking about Rei Toei but rather is discussing some different project that he has also worked on (and also *in a sense* perhaps?):

> Kuwayama, the man with the rimless glasses, was answering something Rez has asked, though Laney hadn't been able to catch the question itself. "—the result of an array of elaborate constructs that we refer to as 'desiring machines.'" Rez's green eyes, bright and attentive. "Not in any literal sense," Kuwayama continued, "but please envision *aggregates of subjective desire*. It was decided that the modular array would ideally constitute an architecture of articulated longing . . . " The man's voice was beautifully modulated, his English accented in a way that Laney found impossible to place.[101]

The more I read this passage, the more I appreciate what Gibson is doing here. His text simultaneously teases us with hints of Rei Toei's artifactual nature and refuses to commit to what it seems to say. Take that gobbledygook about the "aggregates of subjective desire." Let's say we decide that it is all about the Idoru (it is easy for us to do so because Gibson has primed us to feel curious about her origins and nature). The word "aggregates" seems to tip the scales, if ever so gently, toward our thinking of Rei Toei as an artifact. At the same time, our assumption that Mr. Kuwayama is talking about his protégée is our own and not terribly plausible. Since this is a social occasion involving a young woman, her fiancé, and their friends and acquaintances, what are the chances that Mr. Kuwayama would choose to talk in front of them all about the constitution of the young woman's "longing" brain? Social conventions do not go out of the window simply because one of the dinner participants happens to be a holographic projection. It is just as likely, in other words, that Mr. Kuwayama is talking about something else, and the phrase about "aggregates of subjective desire" does not refer to Rei Toei.

Moreover, a different possibility presents itself. Mr. Kuwayama might be speculating about human mentality from a particular AI angle. As practitioners of cognitive science know, the vocabularies describing human cognition and artificial intelligence frequently overlap and will do so even more in the future. In particular, the language of modularity, so frequent in AI discourse, is commonly used among cognitive psychologists today to characterize our evolved cognitive makeup.[102] It is quite appropriate, then, that, next to the reference to "desiring machines," we get the glimpse of Rez's "green eyes, bright and attentive." Rez on this occasion is a compelling illustration of what it means to be a "desiring machine"—in the best sense of the phrase. And if Rez is playfully thought of as a "desiring machine," then what about Mr. Kuwayama himself, a brilliant engineer, whose voice is "beautifully modulated" (and thus reminiscent of that of a well-built animatron)? Who is a machine here, and what does it mean to be a machine, and why are we to assume that there is something "wrong" with bright, attentive, and creative desiring machines?

My point is not that Gibson highlights the machine aspects of the human. Of course he does, and so does everybody else who writes fiction—this is an old trick and pointing it out would not constitute any critical revelation. (What *is* revealing is the cognitive patterns behind it, but I will leave off the discussion of those until later in this book.) I merely observe that Gibson does not commit to a single unqualified reference to Rei Toei as an artifact "built from scratch" with a specific function in mind.

And because Gibson refrains from using the rhetoric of unqualified functionalism to describe his Idoru, we are allowed to take the story of her subsequent development as evidence of her humanlike complexity. This is an important point because, once "born," both Phyllis and Rei Toei do the same thing. They start learning and, by learning, they transcend their present selves again and again. Phyllis reinvents herself with every new profession, as a stage actress, a movie actress, a movie director, a talk-show hostess, and a politician. Similarly, Rei Toei's "only reality is the realm of ongoing serial creation. . . . [She is] entirely process; infinitely more than the combined sum of her various selves. The platforms sink beneath her, one after another, as she grows denser and more complex."[103] Both "artificial" women thus do what we do once we are born: they learn and change. But—and herein lies the crucial difference in our perception of them—because Phyllis is always described in much stronger terms as a "made" thing, we perceive her ongoing education as somehow more mechanical, less human, and generally inferior to the wonderfully creative kind of education that Rei Toei seems capable of.

Consequently, I am sure that had Rei Toei been "turned off" forever at the end of the story by some selfish underachiever, our grief would have been stronger and more immediate than the grief we feel when Pierce dispatches Phyllis.[104] As it is, we are happy that Rei Toei found her Rez, who indeed has enough "imagination" to appreciate this unusual woman and to envision their future life together somewhere in the hidden nooks and islands of the net. It is as if Bridget Jones were indeed allowed to marry the character Fitzwilliam Darcy as played by Colin Firth and to live with him behind the numerous screens of the editing room. Or (to return to more familiar territory) think of Rez as a Prince Charming who can discern the possibility for a lifelong emotional and intellectual partnership where other men see a mere "software dolly wank toy" and admit to having no "idea what [Rez is] talking about when he says he wants to marry her."[105] By softening the edges of his functionalist rhetoric, Gibson has made it easy for us to assimilate his cyberpunk to our romantic fairy tales. By hardening that rhetoric, Berger has left us perennially unsure whether we should grieve over Phyllis's death or merely regret it as much as we regret the demise of a costly and unusually sophisticated toaster oven.

13. *More Human Than Thou (Piercy's* He, She and It*)*

Let's try to formulate some general "rules" based on the use of functionalist language in *Adventures of the Artificial Woman* and *Idoru* (a recipe of sorts for those of you who are into writing science fiction stories). If a writer wants to portray a robot/android/AI program as more human and less mechanical, he or she should make relatively vague both the story of the robot's origins and the description of its function. The broader the function, the easier it might be for us, multifunctional beings that we are, to identify with such a creature.[106] Moreover, a strong emphasis on the obvious mechanical origins of the artificial being ("I built it from scratch right here in my garage with my Erector Set!") may eventually help the narrator to get away with a murder, for who really cares about the demolition of even the most gussied-up toaster oven?

Of course, other narrative elements present in the story impinge on these "rules" and thus mediate the effects of the rhetorical manipulation of our evolved cognitive predispositions for functionalist and essentialist thinking. For example, to render his holographic heroine more human, Gibson has made it possible for us to assimilate his futuristic tale to a familiar plot of romantic discernment and overcoming: a man falls in love with a woman who until now has not been considered an eligible mate by others and, by doing so, instantly changes and elevates her status and level of desirability. By contrast, Berger has intensified the ontological ambiguity of his animatronic female by refusing to align her story with any of our easily identifiable romantic scenarios. (In my personal view, this makes *Adventures* more interesting than *Idoru:* Berger takes more chances with his reader.)

But, say, after we have taken into consideration all those other narrative complications, can we claim that the strength of functionalist rhetoric still governs the ethics of a given science fiction story to a certain significant degree? That is, can we claim that the more ambiguous either origins or function, or both, of an artificial protagonist are, the more empathy we feel for such a being when she is alive and the more grief we experience when she dies? Let us test this tentative rule using another novel, a much longer one, with multiple subplots, and, moreover, one that pulls and pushes functionalist rhetoric in every imaginable direction, presenting us not with just one artificially made being but with a whole army of made, half-made, self-made, over-made, and made-up creatures. And let us check in particular if what I call "the Berger rule"—"emphasize the mechanical origins and you will get away with murder"—still applies in this case.

The novel is Marge Piercy's *He, She and It.* It is set in the middle of the twenty-first century, after a series of wars and ecological disasters have changed the face of the earth and eradicated our familiar geopolitical alignments. Ostensibly, "It" is a cyborg named Yod, who was built to protect the independent Jewish town Tikva from being taken over by hostile corporations vying for world power. The "He" and "She" of the title are more ambiguous. These personal pronouns might refer to the "parents" of Yod—the AI scientist Avram, who built the cyborg (and whom Yod calls "Father," to Avram's displeasure), and another AI scientist, Malkah, who stepped in during the last stages of the work on Yod and thus contributed crucially to the success of the project. In an ironic half glance to the biblical story of Abraham and Sarah, Malkah might be too old to give birth to an actual baby, but not too old to participate in creating a cyborg and later making love to it to "test" the success of her programming (which she does and deems the programming successful).

Alternatively, the "She" of the title might refer to Malkah's granddaughter, Shira. A talented AI engineer herself, she comes down to Tikva to work with Avram on "socializing" Yod after her marriage falls apart and her ex-husband, Josh, gets custody of their only child, Ari. Shortly before Shira's arrival, Malkah gently breaks off with Yod, and he is then free to fall in love with and form a relationship with Shira, who is first completely put off by the idea of making love to a cyborg, but gradually comes to see Yod as "her better, dearer than human lover."[107] If Shira is the "She" of the title, then "He" may be taken to mean either Yod himself (who is increasingly viewed by the people in the novel as a person in his own right), or Avram, or even Avram's son, Gadi. Shira's first love, Gadi is now a creator of famous "stimmies" (movies of the future, in which you not only watch the action but also directly experience the feelings of the actors and actresses).

Finally, the "He," "She," and "It" might refer not to specific people in the story—or *not only* to them—but also to our familiar differentiations between animated and nonanimated entities. Understood this way, "He, She and It" is a fitting tag for a novel that makes a point of teasing the boundaries between living beings and artifacts of whatever gender. Piercy's characters range from people to household appliances, with all possible permutations of "born" and "made" in between. If I arrange the entities populating the novel from the "most" human to the "least" human, this lineup would look something like this.

First among the most human comes Shira, whose body is almost completely free of biotechnology. All she has is a retinal clock built into her cornea, which gives her a time readout whenever she needs it, and a "plug set into [her] skull to interface with a computer." (People who work with AI in Norika—the name of this brave new world—have such plugs as a matter of fact; some of them also have a "set of jacks" built into their wrists.) Commenting on Shira's lack of bodily enhancements, one character observes that she is "pretty much the way her mama made her."[108] (Note, though, that the word "made" slipped into that paean to naturalness.)

Next to Shira stands Gadi, her former lover, who is "fancied up" with much "cutting and pasting" to embody the latest fashion. Indeed, given Gadi's career and fame, people generally "expect him to look like a polished artifact."[109] Still, his are mainly cosmetic changes—nothing "essential"—except that one of his kidneys had to be rebuilt after a life-threatening incident.

Avram and Malkah have been subjected to more serious technological interventions. Avram has an artificial heart (an important detail to which I return later), and Malkah is equipped with a "subcutaneous unit that monitors and corrects blood pressure." Half her teeth "are regrown" and her "eyes have been rebuilt twice."[110] Still, her eyesight is fading, and at the end of the story, she journeys to what used to be Israel to undergo a series of new surgeries that will replace her failing organs and thus allow her to continue her work—her exuberantly creative AI programming. As she herself puts it cheerfully, "I am a project under development. . . . I will be augmented. I'm an old house about to be remodeled"—language that we have seen applied primarily to robots in other novels.[111]

Next comes Malkah's daughter and Shira's mother, Riva, an "information pirate," whose dedication to raiding corporations' databases and freeing the information that they prefer to hoard has landed her on Norika's list of most wanted criminals. Riva is "extremely augmented." Because of her dangerous job, she has "considerable internal circuitry for combat and communication."[112] (Curiously, Riva tends to emphasize the functionalist rhetoric whenever she speaks of herself and others. For example, she views herself as a tool of a future to come, a future, that is, in which the world's biggest corporations will not exclude the poorest parts of the population from access to vitally important information, as they are currently able to.)

Riva's "considerable internal circuitry" still pales in comparison with that of her lover, Nili. At the first meeting between her family and Nili, Riva introduces the muscled-up beauty as "my darling and well-made bomb," and Nili herself announces that she is "an assassin" who is "here to serve" Riva. The "bomb" part later turns out to be a joke, though Nili's technological enhancements indeed make her capable of physical feats not available to human beings. After observing her exercise routine, Shira wonders if Nili "could be a cyborg," and Yod immediately recognizes her as a congenial entity, "part machine and part human." At the same time, Piercy takes pains to emphasize Nili's "organicity" by pointing out (and on several occasions, too, in case we miss it the first or second time) that after her superhuman calisthenics, Nili sweats, and how! Her "exercise garb [is] soaked"; she "reek[s]"; she smells "like a horse."[113] Intuitively, but with absolute assurance, Piercy thus exploits our essentialist tendencies. We associate sweating with living beings (though with some of them more than with others: mice sweat, fish don't, people certainly do). Because it is "in the nature of" people to sweat, Nili's massive perspiration signals her "essential" humanity. (I am sure that you remember it, but, just in case, let me stress again that here and throughout my argument I use the terms "in the nature of" and "essential" to refer not to some real "essence"—which does not exist—but to our evolved cognitive tendency to think of certain entities as having essences.)

Technically speaking, right next to Nili I should put the creatures called "security apes." These are "people altered chemically and surgically and by special implants for inhuman strengths and speed." They move "heavily as robots, although robots were forbidden to be made in human form since the cyber-riots" and are used as bodyguards and lower-rank policemen.[114] I have trouble, however, with finding a proper spot for these "apes" in my hierarchy. Putting them after Nili and before Yod implies that they are more human than Yod, but because of Piercy's manipulation of my functionalism and essentialism, I tend to perceive these apes as more mechanical than Yod (e.g., they are not shown to have many emotions whereas Yod is overemotional, and they fulfill their function of guarding with a pointed lack of creativity). I will place them here, anyway, but the very fact that I feel "in my guts" that this placement is not correct but cannot think of a better one shows that Piercy has succeeded in creating beings whose ontology is profoundly and irresolvably ambiguous.

In the dead center of our lineup stands Yod, a plausible candidate for the "he" and the "it" of the story. Yod, as Avram is careful to point out, is "not a robot." He is a cyborg—"a mix of biological and machine components." As a machine, he has a rigidly defined function—he is a weapon "programmed to protect us—our town, its inhabitants, our [computer] Base. That's his primary duty." At the same time, protection is a very complex function, and to perform it well, one cannot afford to be socially "naïve and awkward."[115] Yod thus has to be much more than a machine, which means that he is equipped with powerful learning capacities, including those for how to interact socially.

In fact, Yod's specifications spectacularly exceed his function. He seems to be endowed with that elusive quality that we call "free will," since he is capable of deceiving his creator and disobeying his orders, which startles people when they come in contact with him, since "a robot could not disobey." Moreover, thanks to Malkah's creative programming, Yod is what we may call a "sensitive" man, every woman's dream mate. He is programmed to love knowledge—which extends to emotional and personal knowledge—and to seek emotional connections.[116] Toward the end of the story, all that this cyborg who was made to be a weapon truly wants is to live peacefully in Tikva with Shira and be a good father to her son, Ari.

Yod's ambiguous ontological status (is he a man or a machine?) is the central problem of the novel. Every parallel between a person and an artifact (as in the case of Malkah who is about to be "augmented"), every reference to an individual being a "tool" (e.g., Riva) or being controlled by another person or serving another person (e.g., Nili), as well as every mention of an artifact made to resemble a living creature (e.g., a toy animal owed by Shira's son) reflects directly back onto Yod, adding to our reservoir of doubts about what this cyborg "really" ("essentially!") is.

Sharing with Yod the spot in the center of our lineup is the golem—Yod's early-seventeenth-century counterpart. According to Malkah's family legend, her genealogy runs all the way back to the famous Rabbi Judah Loew, who built the golem to save the Jews of Prague from an impending pogrom. We get the story of the golem in short installments interspersed with the main narrative. Malkah is "weaving" this yarn for Yod during their twilight meetings in cyberspace: a myth of origins for a high-strung, emotional machine, a bedtime tale for a child who never sleeps.

Built of clay and animated through the rites of Kabbalah, Joseph the Golem provokes widely different responses in the inhabitants of the ghetto. His creator, Rabbi Loew, thinks of him as a necessary evil: the ghetto certainly needs protection, but Joseph is a "clumsy and dangerous tool that must be carefully controlled" while he is alive and turned back to inanimate clay (by reversing the magic that brought him to life) when his function is fulfilled.[117] Other people, unaware of his inhuman origins (or, in one case, aware but not caring about it anymore), come to like Joseph. The rabbi's favorite granddaughter, Chava, who knows that Joseph is a golem, thinks of him as her best friend; and, after his heroic performance during the attack on the ghetto, he is regularly approached by matchmakers who tell him that, though a poor man, he can now marry any woman he chooses.

Next to Yod and Joseph, we have a whole series of Yod's predecessors, cyborgs starting from Alef, Bet, Gimmel, and so forth all the way to the tenth letter of the Hebrew alphabet, yod.[118] Each of the earlier cyborgs is a step back from Yod's "humanness"—each has fewer "biological" components and is less flexible, less emotional, more predictable, more robotic, more inexorably defined by its original function and thus less able to actually fulfill it. All of them were deemed failures and dismantled by Avram, a family history that makes Yod understandably anxious about his own fate. As he explains to Shira, "If your mother had killed eight siblings of yours before your birth because they didn't measure up to her ideas of what she wanted, wouldn't you be alarmed?"[119] (Yod is rather fond of thinking of himself and his dead "siblings" in pointedly human terms—another pull on our essentialist heartstrings.)

After Yod and his robotic brethren comes the "house." Every well-to-do house in Norika is equipped with a sophisticated computer, whose tasks range from regulating temperature and passing messages between members of the family to singing and talking to children, disarming intruders, and generally protecting the owners. Malkah being a brilliant AI engineer, her house is, of course, additionally enhanced. It has a personality of its own and is capable, among other things, of feeling resentment and jealousy. Specifically, the house considers Yod "an irrational invention" and "a computer who pretends to be a biological life form," comments that make Shira observe to the house that it has become "so judgmental lately."[120]

(Now, see, I could have put the "security apes" right here, among the robots, and in fact, I want to. I can't, however, because they are "originally people," a phrase that sways my essentializing mind in one direction, even as the apes' robotic behavior pulls strongly on my functionalist strings.

Where would *you* put the apes?)

Next to the judgmental house are various robots, who have "only enough intelligence to be programmed for simple functions: cleaning, repairing, mining, manufacturing."[121] These are straightforward artifacts with no ontological ambiguity to them. Or almost none. Just as Piercy makes sure that the rhetoric of functionalism creeps into her description of the "most" human heroine of the novel (when we learn that Shira was "made" by her mama), so she rarely fails to insert an anthropomorphic note or two into her description of everyday gadgets.

For example, at one point Shira decides to get a new hairdo. She finds a "robot-hairdresser on the upper shelf where she had shoved it" earlier that year. The detail about shoving the robot onto the upper shelf signals to us that this contraption is small and portable; no images of full-size mechanical ladies with shampoo shelves built into their bosoms, please. At the same time, once Shira inserts her wet hair into the gadget and dials a program for a certain look, she has to sit and wait "impatiently while the many little hands [do their work]."[122] Our essentialist tendencies are thus given a tiny twinge. Because only primates have hands in our world (even though in English the word "hand" is also used to designate a rotating pointer on the "face" of a mechanical clock or watch), when we hear of creatures with hands, we think of living beings. Like Piercy's other brief, almost throwaway, manipulations of our cognitive predispositions, this one goes a long way in a novel that cultivates a constant state of ontological vertigo in its reader: creatures with "tiny hands" cannot be *completely* mechanical, can they?

My lineup from the "most" human to the "least" human is now complete. Note something peculiar about it. You may agree or disagree with my rating of Piercy's characters' comparative "humanness" (security apes being one obvious bone of contention), but it is likely that you do not consider my whole enterprise misguided and meaningless to begin with. That is, you agree that *in principle* these characters can be arranged in some sort of hierarchy, from more human to less human. After all, one can say that I have simply made explicit an implicit hierarchizing that goes on all the time as we read the novel and attach relative ethical significance to certain events, for example, the killing/demise of a human being versus that of a security ape, versus that of a robot-hairdresser.

And yet on some level this hierarchy *is* misguided and meaningless. For were we to articulate the "rules" according to which a security ape is less human than, say, Malkah, we would find ourselves, once more, in a position similar to that of Sosia who tries to explain to his laughing audience what makes him Sosia, and to that of Colin Firth who lists the features that differentiate one of his characters (Paul) from another (Mr. Darcy). And as I have pointed out earlier, we ourselves would not really fare any better in their situations. We "know" that Sosia is Sosia and that Paul is different from Mr. Darcy, but were we to try to explain why, we would end up sounding ludicrous because anything we might say would be inadequate and superfluous at the same time.

And so in the case of a security ape and Malkah, we might say, well, Malkah is more human because, unlike a security ape, she has not been "altered chemically and surgically" and modified with "special implants." And then we will go, oops, no, actually she has been modified quite a bit and is looking forward to being "augmented" even more. Then we can say, well, it is a really a question of degree. The security apes have been modified to a much higher degree than Malkah, and that has made them less human than she is. And then of course, we would have to face the impossible-to-answer-intelligently question: where exactly is the cutoff point after which a modification turns a seriously-augmented-but-still-human being (such as Malkah) into a "less human" being (such as a security ape)?

And, by the same token, where exactly is the cutoff point that makes "the part machine and part human" Nili more human than Yod? Oh, she was *born* human and then modified to have some machine elements, whereas Yod was *made* (functionalist biases to the rescue!). Besides, Nili sweats and Yod does not (essentialist biases to the rescue!). But security apes were also born human and then modified into humanoid machines. By this logic, they should be perceived as more human than Yod, and yet they are not. Is it because we never see or hear them expressing any emotions, whereas Yod is overemotional? But Malkah's house seems to express quite a bit of emotions (it is judgmental, jealous, and resentful), and yet we do not rank its "humanness" above that of security apes.

There is a good reason that in trying to answer such questions we have to fall back onto the dubious authority of our "gut" feelings. We can't come up with the clear definition of the cutoff point between the "still-human" and "not-quite-human-anymore" because our hierarchization of Piercy's characters is a direct function of our essentialism (as is our certainty that Sosia *is* Sosia and that Paul is *not* Mr. Darcy). Our essentialism, however, reflects our cognitive biases and not the actual existence of a certain essential "humanness" presumably present to a higher degree in some entities and to a lower degree in others (as discussed above in part 2, section 6, there is nothing "essential," for example, about our species that separates us from our close primate relatives).[123] Hence the definition of the cutoff point becomes subject to either arbitrary fixed rules or to endless recontextualizing. For example, "Well, you are still human even if such and such organs are artificial, but if you start showing fewer emotions as well, it means that those artificial organs have finally 'gotten' to you, and you have become 'less' human."

Moreover, the fact that this "essential humanness" is a cognitive illusion makes possible the game that the novel plays with its readers. Piercy counts on (so to speak, for authors do such things intuitively) our readiness to construct hierarchies based on some perceived essential qualities of her characters. She can then throw in tidbits that undercut these hierarchies again and again. For example, Yod is a cyborg, which means he ranks lower in his "essential humanness" than Shira's ex-husband, Josh. But then Piercy has Shira observe on many different occasions how robotic Josh seems compared to Yod, how uniquely emotionally attuned Yod is to Shira, something that her "cybernerd" ex-husband never managed.[124] Josh, Shira feels, seems to come from a different species, the implication being that Shira and Yod belong to the same species, and that Yod is more human than Josh.

Note that such observations would not be as titillating to the reader as they are if essences really existed. To clarify this point, think of some objective, measurable quality, such as height. If you are 6′1″, and I am 4′2″, I can wear high heels, muster an attitude that would allow me to look down on you, and mention frequently that I am the taller person. None of these would make anybody around us doubt that I am shorter than you, whatever else they will think about us. By contrast, when we deal with degrees of humanness, Shira's frequent comparisons of her two partners, in which Josh looks robotic and Yod human, make us wonder if on some deep level (which we can't even fathom with our limited knowledge of AI) and in some ungraspable "essential" way, Yod is indeed more human than Josh.

Bringing in cognitive research in essentialism thus helps us understand how *He, She and It* affects its reader, at least in one crucial respect. Piercy's novel abounds in vignettes in which people are likened to artifacts and artifacts to people. It should be clear now that such vignettes rely on our tendency to think of artifacts in terms of their functions and of living beings in terms of their essences, but this is only half of the story. The other, more interesting half is that these vignettes constantly tease and worry the hierarchy of humanness at the core of the novel. The process of reading *He, She and It* thus could be described as grappling with the essentialist hierarchies that the narrative pulls and pushes in all directions.

Let us look at several of those vignettes and consider their pull-and-push effects. Note that because of the way we reason and write I have to present these effects sequentially—that is, first the narrative pushes us this way, then it pulls us that way—but of course, this is not what happens when we actually read this novel. We are being pushed and pulled often at the same time, responding simultaneously to several conflicting "messages" Piercy's rhetoric of functionalism and essentialism sends us. (Of course, given our individual histories, some of these messages will resonate more strongly with some readers than others, prompting different personal and literary associations.)

In the tale that Malkah tells to Yod, Joseph the Golem occasionally wishes to forsake his role as the defender of the ghetto and to defect altogether. As Malkah puts it,

> [It] occurs to him that he could run off from his fate and live as other men do, as it has occurred to Jews in every time to sneak out of being Jewish, to take on the coloration and the jargon of the prevailing culture—Christian, Islamic, corporate—and simply give up the prickly destiny, the treasure that so often kills.[125]

As it happens, the golem does not run off; instead he submits to his fate and fulfills his mission. Both his erstwhile ambivalence and his ultimate submission can be read in several different ways. On the one hand, the golem's ability to contemplate defection in flagrant defiance of his maker shows that he is a creature with free will and an independent mind—so much more than a sophisticated artifact!—an interpretation that undercuts our hierarchy of humanness. On the other hand, since in the final count the golem does not act upon this impulse, we can read his subversive thoughts as a self-deception of a complex machine that cannot transcend its function even as it fondly imagines that it can. The illusion of free will makes for a happier robot—and the golem is thus reinstated in our hierarchy as a rather less human and more artificial being.

Yet staying your course after you have been tempted to forsake it is not necessarily a sign of more insidious and thus more effective "programming."[126] On the contrary, we are familiar with, and generally find compelling, the argument that the availability and awareness of an alternative choice makes more valuable the person's decision not to make that choice. As Milton puts it famously in his *Areopagitica* (1644), using rhetoric particularly relevant to our present argument, "When God gave [Adam] reason, he gave him freedom to choose, for his reason is but choosing; he had been else a meer artificial Adam, such as Adam as he is in the motions."[127]

The screw is thus turned once more: the golem's decision to remain Rabbi Loew's tool, the old man's weapon against the pogromists, is strongly indicative of the golem's "freedom to choose" and thus aligns him not just with people but, in fact, with those people who make the most informed and challenging decisions, those who are aware of easier paths but choose the difficult ones. Our hierarchy is undercut once more—Joseph certainly seems to belong with such people.

Of course, no matter how frequently and compellingly Joseph gravitates toward the human end of the spectrum, he cannot remain there permanently. Neither can he settle comfortably at the opposite end, however insistent Rabbi Loew might be in calling Joseph a "tool." The golem remains what Boyer has called a "counterontological" entity: he can never be fully assimilated either to the domain of living beings or to the domain of artifacts. His cognitively ambiguous position in no-man's- and no-artifacts'-land is what makes possible his constant fluctuations along the scale of "humanness" constructed by the story. And vice versa: such fluctuations reinforce his status of a counterontological being (a feedback loop similar to that which I have already considered in the case of Asimov's Andrew).

Here is another vignette (or rather a series of vignettes) undermining the hierarchy of humanness of *He, She and It.* This one involves Yod's creator, Avram. I mentioned earlier that Avram has an artificial heart (his own must have been too weak and failed him at one point). The choice of heart as one organ to be replaced in Avram's otherwise "organic" body is significant. To show why, let me first quote at length from Gelman's *The Essential Child:*

> Explicit essentialist assumptions given by adults characterize essences as invisible, distinct from outward appearances, and remarkably stable and resilient. I illustrate this point with the self-reported experiences of Claire Sylvia, who underwent a heart-lung transplant. . . . Following the surgery, she described changes in her behavior and emotions. For example, she began to crave beer and fried chicken—foods she never liked before. She became more aggressive, independent, and confident, and for the first time walked with a swagger. Even her favorite colors changed, from pink and red to blue and green. Sylvia attributed these changes to characteristics of her donor, a young man named Tim. She said she felt "as if a second soul were sharing my body," . . . and referred to "this new male energy." . . . In speculating on what remained of the man who had donated his heart, she concluded, "Perhaps what still existed of Tim was his purer essence." . . . Sylvia's therapist concurs: "I am beginning to believe that some of Tim's essence has transmigrated to Claire. . . . If the transplant has somehow passed on elements of his temperament, personality, and identity, then psychological residues of the actual Tim L. (not just the image of "Tim") may now inhabit Claire." . . . Others that Sylvia interviewed used very similar language: "I could feel his essence, his energy" (Harriet, wife of a heart donor, talking about meeting the recipient of her dead husband's heart . . .). "Anyone who receives a new heart is getting a big ball of subtle energy. Ancient cultures have known about subtle energy throughout history, and have viewed it as the vital force of all creation" (Paul Pearsall, author).

One last example comes from "the wife of heart-transplant recipient, discussing changes her husband experienced after receiving his new heart" (in Claire Sylvia and William Novak, *A Change of Heart,* cited by Gelman): "Of course he's different. There are genes and energy in him from someone else's body. Those things affect you."[128] Scientifically indefensible as these statements might be, they obviously strike a chord in many of us. By saying that they strike a chord I do not mean that if somebody else's heart is transplanted into my chest, I will necessarily think that the transplant now impacts my personality in ways subtly resonant with the personality of its original bearer. In fact, as far as I know myself, I don't think I can take seriously the idea that the new heart would carry personality traits of the donor.[129] What it rather means is that I find this kind of thinking *plausible.* To make this clearer, consider the following example:

> The donor was an inveterate gambler. You hate gambling, but after you have his heart transplanted into your chest, you find yourself more willing to take chances. You wonder if some of his gambling "nature" has influenced your personality, if ever so slightly.

This is to say that even if we consider a joke the idea that someone's personality can take on some of the features of the personality of her heart donor, this idea still makes sense. That is, we do not need any explanation of why this kind of "transference" might be possible and why some people might believe in it. My point then is that the reason we don't need this explanation is that this idea taps our essentialist biases.

To explain the heart's privileged status as one of the "essence-preserving" organs or substances (others include brain and blood), we can say that it has always occupied a special place in at least the Western cultural imagination (see, for example, Robert A. Erickson's *The Language of the Heart, 1600–1750* [1997]). To approach the same issue from a slightly different angle would be to ask how the heart came to occupy this special place. Perhaps, essentializing species that ours is, our cognitive architecture is always "on the lookout" for an organ or substance that could be invested with this kind of meaning. The heart represents one convenient embodiment of our essentialist intuitions because of its placement (deep inside our chest; invisible for all practical purposes), its vital importance (we die when it fails), and some semblance of agency associated with it (we can actually hear its "doing" something, and we know it continues doing it even when we are asleep). Thus we can say that although heart transplants are a very recent phenomenon, our thinking about their implications is subject to the same cognitive biases that traditionally contributed to making the heart an essence-bearing organ.

Back to Avram. One striking thing about this brilliant AI designer is that he seems to have very little or no need for personal emotional attachments. We know that he loved his wife (Gadi's mother), Sarah, passionately, but she died a long time ago. Piercy does not specify when Avram got his artificial heart, but it is easy to imagine that it happened around Sarah's death. Avram's relationship with Gadi has always been troubled. He does not seem to like his son, ostensibly because Gadi, formerly a bad student and now a mischievous sort, has never lived up to his father's expectations. Avram's relationship with his friend and occasional colleague Malkah is marred by bickering and distrust. Before Avram met Sarah, he used to be romantically involved with Malkah but that shared past seems to have no bearing upon his present raspy attitude toward her.

Avram lives for his work, but he is not the least sentimental about the cyborgs that he creates and "scraps" when they fail to live up to his vision. Toward the end of the story, he decides that Yod has to die in order to fulfill his function of protecting Tikva. Avram plans to send Yod to negotiate with the top functionaries of the corporation that is trying to destroy Tikva (and that is very interested in acquiring Yod) and to blow him up in the middle of that meeting, so that he takes with him as many of Tikva's enemies as possible. The language that he uses to announce this news to Yod, Shira, and Malkah is downright "heartless." While Shira, "cold with horror," is "looking only at Yod," Avram is explaining:

> I made him, and I can unmake him. This an opportunity to deal an amazing blow to [the corporation]. Yod was created to protect and to defend us. . . . An attack on [the corporation] is absolutely essential for our survival. . . . If we don't show we can hurt them back by assassination for assassination, we're doomed. [Moreover,] I can manufacture another [cyborg. With money from the corporation], I can manufacture another exactly the same, starting tomorrow. . . . I can improve on the design. Kaf [the next cyborg] will be superior.[130]

Avram knows that Shira loves Yod and that Malkah is also very attached to him. It is not that he wants to hurt either woman, but he simply does not seem to understand what effect his words will have on his listeners. In fact, he is surprised by the strength of their negative reaction. As he puts it, "I created Yod, and indeed, I seem to be the only one who remembers his purpose." As to Yod, Avram appears to think that he would actually be glad to know that after his demise, another, "superior" cyborg will take his place. "There will be more of you, I promise," he tells Yod, who is clearly not thrilled by the perspective.[131]

Of course we can always read some poignant emotions into Avram's outwardly cool behavior (and compare his actions to Abraham's sacrifice of Isaac). We can insist that he is actually quite attached to Yod. "Designed to perform well at activities Gadi failed," Yod is the perfect son Avram never had as well as the pinnacle of his scientific ambitions.[132] This would mean that Avram is willing to sacrifice for the good of his town one thing that he truly loves and that he refuses to acknowledge his feelings even to himself because it may make him protect Yod and thus further endanger Tikva. It is almost as if the person who most needs to be persuaded that Yod is a machine designed for a certain purpose is Avram himself, hence his unrelenting rhetoric of functionalism around Yod.

This interpretation is not impossible—for which father, even the most cold-hearted, can we say deep down does not really love his child?—but we do have to work rather hard to convince ourselves of its truth. I remember that on my first reading of *He, She and It,* I positively hated Avram for his decision to sacrifice Yod; I considered him a monster and wished he had died before he put his sadistic plan to kill Yod into execution. I am not sure that I like him more today even after I have come up with this appealing he-feels-more-than-he-shows theory.

It seems to me that Piercy slips in the mention of Avram's artificial heart early in the novel to introduce the theme of his emotional aloofness and to prepare us for his cruel behavior at the end of the story. If this is the case, and I strongly suspect that it is, then she is intuitively appealing to the same essentialist biases that prompt the people quoted in *A Change of Heart* to believe that the heart indeed preserves some "essential" qualities of its original bearer and that those qualities can manifest themselves later in the heart's new owner. To the questions of what kind of person would send Yod to die and what kind of person would remain oblivious to the pain that this decision is inflicting on Yod, Shira, and Malkah, we can thus say, a person who no longer has a "human" heart in his breast and whose humanness has deteriorated in response to this infusion of the artificial into the "core" of his being. Avram's position in our hierarchy of humanness is thus given a shaking. It is not that we decisively move him in the direction of the "artifact" end of the hierarchy, but we do begin to feel that his position on the "human" end is worth checking and reweighing now and then.

Let us now see if the "Berger rule"—"emphasize the mechanical origins and you will get away with murder"—still works in the case of Avram and Yod. On one level, it does. When Avram explains his plan for sacrificing Yod, we may hate him for his heartlessness, but we do not absolutely condemn him for not doing everything in his power to avoid sending Yod to commit a suicide attack (as he would have tried, we assume, in the case of a human being). Does this mean that the functionalist rhetoric—mouthed tirelessly by Avram who calls Yod an "object" and insists that Yod was "manufactured for a purpose still unfulfilled"—is somewhat effective?[133]

It *is* effective, I suggest, to the extent to which readers buy into Avram's reasoning. It is difficult to foretell how a specific reader would respond to the interplay between Avram's implied literal heartlessness and Yod's objectification. If the former wins out, then Avram might be perceived as an unreliable judge of Yod's status (just as Pierce is of Phyllis's, though, obviously, for different reasons). But if a reader puts less weight on Avram's own status as a man with an "artificial" heart, then she may pay more attention to other factors that seem to support Avram's insistence that Yod must be used according to "its" purpose.

Such a reader may thus experience Yod as more expendable than Shira, Malkah, Gadi, and other human protagonists. What may support this interpretation further is that in the discussion of Yod's intended mission, the narrative seems to focus less on Yod's emotions than those of Shira, who, we know, will be devastated by Yod's death. As such it seems to implicitly accept the view of Yod as a "tool" of somebody else's happiness; what happens to him matters less in terms of what it does to him than what it does to other people. In other words, one of several plausible readings of the novel seems to lend credibility to the "Berger rule." To the extent to which we ignore the source of functionalist rhetoric, Avram (that source) is getting away with murder.

There is yet another complication to the "Berger rule" as borne out by this novel. Avram may be getting away with murder—having persuaded us that forcing a cyborg to commit a suicide attack is not as morally indefensible as forcing a human being to commit a suicide attack—but he does not survive it. To ensure that Avram never builds another "conscious weapon" to protect Tikva, Yod arranges it so that the moment he blows himself up amid Tikva's enemies, Avram's laboratory is destroyed too, and Avram's body is blown to pieces by the powerful explosion.[134] Yod's decision to thus kill his maker has many implications, but what I find particularly interesting for the purpose of the present discussion is a certain feeling of justice well served accompanying the news of Yod's last action. What this means is that even though we may never experience Yod as fully human (the abundance of functionalist rhetoric works hard toward that end), we still on some level conceptualize him as human enough to think of his last action as rightful revenge on Avram (even if Yod himself does not think about it in such terms and claims to like and understand Avram). In other words, the novel's frequent references to Yod's rich emotional world (for remember that it is "in the nature" of human beings to have rich emotional worlds!) may never completely neutralize the rhetoric of functionalism applied to him. Still, it humanizes him to the point that our old notion of taking life for life seems entirely pertinent to what has taken place between him and Avram.

Yod's behavior thus vindicates our Frankenstein complex with a vengeance. I earlier redefined this complex to mean our intuitive desire to resolve the cognitive problem represented by the figure of a "living artifact." When such a conceptual hybrid rebels against its function, it strongly aligns itself with living beings, multifunctional and unpredictable. *And so it should*—we feel—given its apparently rich mental world and other "essential" human features. Yod has been questioning his function and mildly disobeying Avram throughout the story; by putting an end to Avram's production of cyborgs who would presumably carry on his mission of protecting Tikva, he decisively transcends his own function as Tikva's defender. Alternatively, we can say that he rewrites his function, by making sure that the yet unborn cyborgs will never know the mental anguish that being a "conscious weapon" would have put them through. However, if anything, this interpretation reminds us that Yod's last action is now as open to multiple and conflicting readings as anything that people do. Once more, Yod's position in the dead center of our hierarchy of humanness has swayed toward the "very human" end of the spectrum.

In conclusion, let me point out that we certainly do not need the insights from cognitive theory to realize that Piercy's novel makes us question the meaning of the term "human." Nor do we absolutely need cognitive theory to speculate about why we are attracted to the idea of such questioning. What cognitive theory does, however, is "sharpen both the answer to the why question and the discussion itself."[135] Knowing that we tend to essentialize both living beings and abstract concepts helps us to see why concepts such as "human" are open to endless redefinition and recontextualization, as opposed to such concepts as "chair" or "cup" or "rifle," which are largely thought of in terms of their current or intended function with not a whole lot of room for epistemological maneuvering.

What I have presented here is a series of unavoidably limited readings of a complex novel whose characters embody various modes and models of such redefinition. Still, whether you expand my interpretations further or drastically disagree with them and offer your own, you will necessarily be drawing on the host of implicit assumptions grounded in our evolved cognitive categorization biases. We never spell out these assumptions explicitly. Indeed we are not even aware that we make them. How many of you, for example, said to yourselves as you were reading the novel, "Nili sweats; it is in the nature of living beings to sweat; Nili must be a living being, not a robot," or, "Riva says that she is a tool; tools are defined in terms of their function; Riva's actions must be subordinated to whatever she perceives as her function," or, "Yod is jealous of Gadi because he knows that Gadi still has some emotional power over Shira; only people experience such complex emotions; Yod must have something essentially human about him," or, "Avram's real heart has been replaced with an artificial heart; when your heart is replaced, some of the qualities of its original bearer rub off on the new owner; Avram may now exhibit certain features of a robot, such as lack of empathy"? We do not say or even think such things, but it is clear that we would not have been able to make sense of a single page of *He, She and It* nor interpret the novel within any other literary-theoretical framework had we not subconsciously relied on our differential conceptualization of living beings and artifacts as a function of our cognitive evolutionary heritage.

14. Made to Pray

I have considered so far fictional narratives in which an artifact created to serve a certain limited function rebels, whether peacefully or violently, against its creator, transcends its function, and ends up relatively unpredictable and multifunctional.[136] Asimov's Andrew, Seidelman's Mr. Right, Berger's Phyllis, Gibson's Idoru, and Piercy's Yod all do that. But the directly opposite conceptual move is also possible. A human being—somebody we would ordinarily think of in terms of her ineffable personal essence—can be forcefully conceptualized as "made" with a specific function in mind. This happens so often both in fiction and everyday life that we do not even consider it a clever rhetorical gambit. Nevertheless, that's what it is: a rhetorical gambit that derives some of its effectiveness from tapping our cognitive categorization biases.

To see how this works, let us begin with a stark example from children's literature. In 1781, poet, writer, and pedagogue Anna Laetitia Barbauld published the second book in her series dedicated to the religious education of young readers. Entitled *Hymns in Prose for Children,* it featured fifteen vignettes describing various aspects of God's close involvement with the natural and social world. Generously illustrated and occasionally catechistic, *Hymns* must have struck a chord with parents and educators.[137] It went through multiple editions over the next two centuries, the latest edition coming out as recently as 1996.[138]

Barbauld's book opens with the following paraphrase of the first chapter of Genesis:

> Come, let us praise God, for He is exceeding great. . . . He made all things; the sun to rule the day, the moon to shine by night. He made the great whale, and the elephant; and the little worm that crawleth on the ground. The little birds sing praises to God, when they warble sweetly in the green shade. The brooks and rivers praise

> God, when they murmur melodiously amongst the green pebbles. I will praise God with my voice; for I may praise Him, though I am but a little child.[139]

Right away, Barbauld introduces the crucial theme of God's relationship with his creatures. God is a maker; the objects of his craftsmanship—"all *things,*" including birds, brooks, and children—praise him in their various ways. The second hymn makes the same point. After a lively description of blooming flowers, fruit trees, and sporting animals (goslings, chickens, lambs, and butterflies), it tells us that all living creatures "thank Him that hath made them alive."[140] In the third hymn Barbauld shows us a rose, a lion, and the sun and then turns again to him who "made" (a word repeated five times in the course of this 342-word hymn) the rose, the lion, and the sun. Hymn seven repeats the theme of God the maker: "We can praise the great God who made us"; "we that are so young are but lately made alive"; "He fashioneth our tender limbs, and causes them to grow; He maketh us strong, and tall, and nimble"; and finally the phrase that I see as a leitmotif of this cluster of hymns, "man is made to praise God who made him."[141]

Thus in the first seven hymns children are portrayed as engaged or as being encouraged to engage in only one activity: praying and thanking God for making them. God on the other hand is presented as a skillful craftsman responsible for producing a variety of living beings, including human children. Barbauld reiterates here the age-old paradigm describing the ideal relationship between God and his creatures, echoing in particular the passage from Milton's *Paradise Lost* in which Adam tells Eve that they should "ever praise him" that "made us."[142] Of course, Barbauld's intended readers—three- to five-year-olds, untutored in biblical and Miltonic imagery—could not appreciate the illustrious theological and literary pedigree of this sentiment. The appeal of the juxtaposition of the two ideas—God "fashionedth" children; children ought to pray to Him—thus lay somewhere else.

We can argue today, armed with insights from cognitive science, that it lay in its activation of certain properties of our evolved cognitive architecture. Barbauld's image of a child made to praise the God who created this child draws on two cognitive domains: one associated with natural kinds and one associated with artifacts. The child, a living being, is characterized as being "made," artifact-like, by the omnipotent craftsman, eliciting, to quote Frank Keil again, "both essentialist and functionalist interpretations."[143] The effect of such a characterization is that, because artifacts are typically perceived in terms of their intended functions, it becomes easier for readers to conceptualize the child as having a function—in this particular context the function of praying. Or to put it differently, the importance and duty of praying are legitimized through the implicit but cognitively compelling appeal to properties attributed to "made" objects.[144]

Something else perhaps is going on here, too. Recent studies by E. Margaret Evans and Deborah A. Kelemen suggest that children of kindergarten age might be particularly open to thinking of living beings from a certain functionalist angle. In fact, Kelemen considers young children "intuitive theists" because of their "promiscuous teleology," that is, their tendency to apply the reasoning reserved for artifacts (e.g., chairs are for sitting) to natural entities (e.g., clouds are for raining). Young children's cognitive bias toward a teleological (or functionalist) view of living entities manifests itself in their conviction that "natural entities are 'made for something' and that is why they are here."[145]

Moreover, Bering reports that Evans and her colleagues "have found evidence that children prefer creationist arguments over evolutionary ones when reasoning about the origins of species. Teleological reasoning is always applied to the origins of the self, such as talking about what one was 'born to do' or that one is leading a life that he or she was not 'meant for.' . . . This type of . . . teleo-functional reasoning . . . may set the stage for an obligatory social relationship between the self and its presumed supernatural creator."[146] In other words, kindergarten-age children may constitute *the* perfect audience for an argument that praising God and praying to him is what they were "made" for.

In fact, we can correlate the "teleological promiscuity" of kindergarteners with the phase of their cognitive development characterized by an active exploration of the possibilities of cross-mapping between the conceptual domains. Remember the experiments that strongly imply that five-month-olds do not readily perceive people as fully material objects? The more mature realization that at least on some level people *are* material objects has to develop somehow. It is thus possible that the teleological tendencies manifested by children of kindergarten age are an expression of that maturation process. Thinking of people as "made for something" allows the children to push and test the boundaries of their functionalist and essentialist tendencies.

Suggestively, this period of intuitive theism overlaps with the crucial threshold in the development of children's theory of mind. As we remember, prior to the age of four, children are generally unable to attribute false beliefs to others: they assume if they have access to certain information, then other people do too. According to recent studies conducted by H. Clark Barrett and his colleagues, "the social cognitive systems of young children [thus] may be better suited to reasoning about the culturally postulated mind of God than about the epistemologically limited minds of humans and other animals." God, after all, is "omniscient and therefore cannot hold false beliefs (and therefore cannot be deceived)"—a state of mind that is easier to imagine if your own theory of mind is relatively immature and you tend to think that other people always know what you know and thus cannot hold the beliefs that you know to be false. "For example, whilst 3-year-olds *incorrectly* reason that a naïve person knows the true contents of an inaccurately labeled box, they *correctly* reason (at least in a theological sense) that God knows the true contents as well."[147]

This is not to say that children of that age are "naturally predisposed" toward religion as such. Rather, the particular pattern of intense cognitive exploration that accompanies the maturation of various cognitive domains in three- to five-year-olds may render them peculiarly open—or shall we say vulnerable?—to the patterns of representations prevalent in certain religious beliefs.[148]

The studies that show that children may approach living beings from a teleological or functional perspective (e.g., "Why I am here?" "What was I made for?") complement in interesting ways the earlier-quoted studies by Atran and his colleagues. Consider the perceived stability of the function of a given artifact as opposed to the perceived stability of the function of a human being. The child can be easily convinced that a chair is made for sitting, but try convincing her, without changing anything about the chair's appearance, that it is made for toasting bread or for wearing in cold weather. By contrast, a child might be compelled to believe that she was "made" to pray to the God who made her, but she could also be compelled to believe that she was "made" to keep her mommy happy, or that she was "made" to learn as many new things at school as she could, or that she was "made" to play chess. The instability of children's functionalist, or teleological, approach to living beings is thus revealed by the multiplicity and volatility of their possible functions.

To return to Barbauld's children who are "made to pray": the effectiveness of this image's functionalist rhetoric thus stems paradoxically both from the difference between the respective cognitive domains that process information about living beings and artifacts *and* from the adventurousness of our cognitive predispositions. The idea that I was made to serve a particular function taps the domain of artifacts, but it also teases my essentialism (as in, "Is this function indeed my *essence*?"), titillates my theory of mind ("If I was *made* with a certain function in mind, what does it say about the *mind* of my maker?"), and recalibrates my perception of myself as, among other things, a material object.

Barbauld of course did not think in terms of living kinds and artifacts when she wrote *Hymns.* Writers and rhetoricians have never needed cognitive scientists to explain to them that what they do constitutes activating inferences from different conceptual domains in a particularly felicitous fashion. What makes Barbauld's case more interesting in this respect is that she actually thought through the intended rhetorical impact of her writings and developed a "prototheory" of the child's cognitive development. This prototheory is appealing in many ways, but, ironically, it is also based on assumptions that run counter to those held by cognitive evolutionary psychologists and anthropologists today.

Barbauld saw her *Hymns* as radically different from the "multitude of books professedly written for children" and yet "not adapted to the comprehension of a young child."[149] In the preface to her book, she criticizes her famous predecessor Isaac Watts, author of *Divine Songs for the Use of Children* (1715), for his well-intentioned but, in her opinion, misguided project of addressing his young audience in verse. (Barbauld builds here, perhaps, on Rousseau's argument that small children can only be confused by poetic fables.)[150] As she puts it, Watts is "deservedly honoured for the condescension of his Muse, which was very able to take a loftier flight," but, generally, poetry is wasted on young readers. It should not "be lowered to the capacities of children. . . . They should be kept from reading verse till they are able to relish good verse; for the very essence of poetry is an elevation in thought and style above the common standard; and if it wants this character, it wants all that renders it valuable."[151]

Another problem plaguing contemporary religious literature for children is, in Barbauld's view, the unnecessary artfulness of story lines. She argues that a "connected story, however simple, is above [the] capacity . . . of a child from two to three years old" and only interferes with the grand project of impressing upon the child's mind the "full force of the idea of God."[152] It is much more beneficial, she contends, to connect "religion with a variety of sensible objects, with all that [the child] sees, all he hears . . . and thus, by deep, strong, and permanent associations, [lay] the best foundation for practical devotion in future life." This task is "humble, but not mean; for to lay the first stone of a noble building, and to plant the first idea in a human mind, can be no dishonor to any hand."[153]

On the whole, Barbauld's critique of the competing books for children, those burdened with "connected stories" or poetry, was fairly conventional. Lamenting the inadequacy of available religious literature for young readers had long been a tradition among English educators. As early as 1712, William Jole complained in *The Father's Blessing Penn'd for the Instruction of His Children* about what he perceived as the weaknesses of such literature:

Of all the Books in print, I cannot find
One Godly Book exactly to my mind,
Meetly proportioned to Children's strength;
Some are too short, but most do err in length.[154]

As Patricia Demers notes in her analysis, running up to 1850, of moral and religious literature for children, the situation had not seemed to improve much by the close of the eighteenth century. In 1795, Dorothy Kilner observed in her *First Principles of Religion* that "in a cause of such infinite moment as implanting the first principles of religion on the minds of infants, no age has yet exerted themselves, or fixed any rational plan of instruction."[155] A "godly" book, a book that answered the needs of a cause of "such an infinite moment," would be the one that could, in Lady Eleanor Fenn's words in 1783, "catch [children] gently" by adjusting to their conceptual level and rendering them more pliable for further moral instruction.[156]

And catching children gently by adjusting to their capacities is precisely what Barbauld wants to achieve in her *Hymns*. Here is her explanation of what she is doing. Barbauld argues that the idea of God should be stripped of any embellishments and inculcated in the child's mind on the level of immediate perception, so that the child "never remembers the time when he had no such idea." Each "sensible object, all that he sees, all he hears" should be mediated by religious thought, so that he will "see the Creator in the visible appearances of all around him." There is a certain danger associated with it of course, for the child's "religious ideas may be mixed with many improprieties." Still, with time "his correcter reason will refine [those] away," and the child will emerge awash in "that habitual piety, without which religion can scarcely regulate the conduct, and will never warm the heart."[157]

Barbauld here seems to share certain associationist premises with John Locke, even while her practical advice concerning the age at which the child should be exposed to religious ideas diverges from his. As Locke puts it in his *Essay Concerning Human Understanding* (1690), "Let custom from the very early childhood have joined figure and shape to the idea of God, and what absurdities will that mind be liable to about the Deity!"[158] To use modern lingo, Barbauld apparently assumes that young children have no innate cognitive structures for concept formation. Thus if one catches the child at that tender age when he or she is first ready to invest mental representations of external objects with meaning, one can radically influence the structure of these representations or, to use Barbauld's own words, "plant the first idea in a human mind."[159]

From the perspective of cognitive analysis, Barbauld in *Hymns* does the opposite of what she thinks she is doing. In her "Preface" and "Advertisement" she implicitly represents the child's mind as a blank slate liable to take in and to bear any inscriptions, however free ranging and arbitrary, offered by the surrounding culture.[160] At the same time, the key "message" of the first part of her book—children ought to pray to God because he made them—is grounded both in our cognitive tendency to conceptualize living beings differently from artifacts and in our constant readiness to probe the boundary between the two conceptual domains. Barbauld's "message" makes sense on what we may call an intuitive level precisely because her young reader's mind is *not* a blank slate. It is rather an infinitely complex agglomeration of cognitive susceptibilities adapted in the process of human evolution to recognize and interpret environmental stimuli according to the perceived properties of those stimuli. For example, when the properties of a given object (in *Hymns,* the child) seem to satisfy the input conditions of the domain of artifacts (the child was "made"), certain inferences associated with the domain of artifacts (e.g., "artifacts have functions") will be activated, and the reader will be favorably disposed to consider claims about the uniformly defined "function" of this object (here the function of praying).

Consequently it is unlikely that a child would be equally open to associate *any* "absurdity" with the "idea of God." Some "absurdities" are more cognitively felicitous than others and thus are remembered better and picked up by other members of the culture more readily. As cognitive evolutionary scientists Leda Cosmides and John Tooby argue, "The assumption that mental representations with different content are equally easy to transmit is false. Representations whose contents taps into a domain for which we have specialized mechanisms will be transmitted very differently than representations whose content does not tap into such a domain."[161] Independently from the writer's awareness of her own and her readers' cognitive processes, she has to mobilize our domain-specific cognitive architecture in the attempt to influence readers.

15. *Made to Serve. Made to Obey. Made to Break Hearts*

The implicit appeal to evolved cognition thus emerges as a crucial element of a rhetorically compelling argument. In her *Hymns,* Barbauld straddles the cognitive domains that process differentially information about living beings and artifacts in the service of her project of early religious education. Let us consider now other possible applications of such rhetorical straddling, for it seems that the functionalist approach to human beings can be mobilized to support a wide variety of ideological agendas. An argument that begins with the premise that some groups of people—social classes, castes, ethnic groups, sexes—were "made" to perform specific duties derives some of its power from activating our cognitive proclivity to associate "made" objects with rigidly defined functions. Thus Rousseau in *Emile:*

> In the union of the sexes each contributes equally to the common aim, but not in the same way. From this diversity arises the first assignable difference in the moral relations of the two sexes. One ought to be active and strong, the other passive and weak. One must necessarily will and be able; it suffices that the other put up little resistance.
>
> Once this principle is established, it follows that *woman is made specially to please man.* If man ought to please her in turn, it is due to a less direct necessity. His merit is in his power; he pleases by the sole fact of his strength. This is not the law of love, I agree. But it is that of nature, prior to love itself.
>
> If woman *is made to please and to be subjugated,* she ought to make herself agreeable to man. . . .
>
> [D]ependence is a condition natural to women, and thus girls feel themselves *made to obey.* . . .
>
> The first and the most important quality of a woman is gentleness. As *she is made to obey* a being who is so imperfect, often so full of vices, and always so full of defects as man, she ought to learn early to endure even injustice and to bear a husband's wrongs without complaining. . . . *Heaven did not make women* ingratiating and persuasive in order that they become shrewish. It *did not make them weak* in order that they be imperious.[162]

However offensive Rousseau's arguments may feel to us today, we cannot deny that he was a brilliant rhetorician. *Emile* is a rhetorical tour de force that uses every (dirty) trick in the repertoire of a consummated public speaker—emotional manipulation, calculated fluctuations of style, timely mention of scientific "facts," passionate personal appeals, strings of apparently incontrovertible syllogisms, entertaining "real-life" examples, and fictional vignettes—to convince the reader of the truth of its arguments. When I teach *Emile* today, I routinely ask my students to write short papers that imitate stylistic peculiarities of this novel-cum-philosophical treatise. Along with some obvious parodies, I also get essays that strike both its authors and other people in the class as "creepy," for they demonstrate that by closely imitating Rousseau's rhetorical strategies one can render convincing initially outlandish ideas.

This is to say that Rousseau would not choose to sound like a broken record—

repeating three times on the same page (and then three times on another page) that woman "was made" to serve and obey—if he did not know intuitively that such a repetition served his purposes exceptionally well. The writer who spent his life honing his rhetorical skills must have felt that harping on what I call here the "artifactual" nature of women strengthened his argument about the "naturally" limited function of women as compared to that of men, which in turn supported his larger premise that the sexes have a "common aim in their union but . . . different ways of contributing to that aim."[163]

For note how the notion that "woman is made specifically to please man" allows the narrator to differentiate between the inescapable *function* of women and a mere *predisposition* of men. If the woman directs her energy toward something other than "pleasing a man" or if she becomes "shrewish" and "imperious," then she goes against her very nature, rendering herself and everybody around her miserable and violating the Heaven-ordained way of things. (A can opener may "decide" that it won't open cans anymore and break down, but nobody, including that temperamental artifact, will be better off for it.) By contrast, man is not *made* to please woman—he merely "ought to please her in turn." His desire to please is "due to less direct necessity": it is "in his power" to decide whether to exert himself in such a fashion or not. Rousseau's use of functionalist rhetoric to naturalize the second-class citizen position of women in a patriarchal society is thus typical in the sense that when applied to social groups, such rhetoric is used more often than not to justify oppression and discrimination.

Here are some less ideologically laden and more casual uses of such rhetoric (though, in actuality, ideology is rarely completely out of the picture in such cases). We intuitively rely on the language of functionalism in our everyday discourse when we want to emphasize the superlative degree to which a person possesses a certain quality. For example, when we describe two people as "made for each other," this phrase conveys a stronger image of a mutual fit than the ostensibly similar phrase "they are a perfect couple." If we want to praise somebody's performance, "she is born to dance" feels more rhetorically compelling than "she is a great dancer." "The best killers" sounds extravagant enough, but it is still weak tea compared to "natural-born killers," for the latter builds on the tacitly functionalist "born to kill." Here is why this functionalist rhetoric can be so powerful: our cognitive evolutionary heritage apparently leaves us "no choice" but to process at least some of the information about the human beings thus described through the cognitive domain of artifacts.

It so happens that my own favorite cases of the use of such rhetoric in fiction have to do with the relationship between parents and children, and I am grateful to Mark Turner, who first pointed out to me how readily parents start speaking the language of functionalism when they want to manipulate their offspring, or, at least, want to alert them to their duties to the people who brought them into this world. At first glance, statements such as, "I gave you birth; you should pay at least some attention to what I am saying," are emotionally intelligible because they seem to imply the indebtedness of the child to the parent: "I did something so huge for you, something you can't ever hope to repay me for, that the least you can do toward discharging some of that looming debt is to treat me nicely now." This "economist" reading tells only one half of the story, however. The other half has to do with the logic of functionalism underlying such sentiments. Because I gave birth to you—brought you into this world—made you—created you—I just might have a privileged perspective on what you should or should not do with your life, *just as a craftsman who creates an artifact certainly has a better perspective on what the artifact was designed to do than the artifact itself.* Although we never think along these lines during actual conversations with our children, reminding them that we brought them into existence must be profoundly informed by functionalist reasoning. For otherwise, just how would either we or they see any causal connection between "I brought you into this world" and "I know what is best for you"?

In the case of fictional families, there is, for example, a Russian catchphrase made famous by the fierce Cossack protagonist of Nikolai Gogol's historical novel *Taras Bulba* (1835), which relies on functionalist rhetoric to justify filicide. The tough old Taras can see only one life purpose for his sons, Ostap and Andrii, which is to fight the enemies of the Cossacks, who at that historical juncture happened to be the Poles. To his wife's meek request that the boys stay home a while, Taras responds categorically that "A Cossack is not born to run around after women."[164]

So when later in the novel, the younger son, Andrii, does throw his lot in with a woman, falling in love with a daughter of a rich Polish nobleman and switching over to fight on the side of her kinsmen, Taras finds him and kills him on the spot with the laconic, "I gave you life, I will also kill you."[165] As a murder, Taras's action can be justified by Andrii's betrayal of his compatriots, but as a cold-blooded filicide it needs a strong additional defense. So I find it interesting that the rhetoric of functionalism offers us here precisely that: an "excuse" for the "unnatural" act of a father's murdering of his son. Taras's choice of words forces us to agree—however much we may dislike it—that on some level his unnatural deed makes sense. To some extent, the young man is momentarily perceived as an object and his father as a maker of that object, even if we never explicitly articulate it to ourselves this way. It is tacitly implied that Taras has the same power of life and death over Andrii as the craftsman has over his artifact. This appeal to functionalism apparently serves to quell (some of) our moral outrage.

Here is another narrative that focuses on relationships between parents and children and plays with our intuitive assumptions about the hierarchy of power between craftsmen and the objects that they bring into existence. Charles Dickens's *Great Expectations* (1861) features several headstrong adults and their adopted children whom they "make" and "mould" according to their darling wishes. And although in at least one case (that of Estella) the child grows up to fulfill the function envisioned for her by the ambitious parent, at the end of the day, the parent suffers precisely because she succeeds so spectacularly with her project of treating a child as an artifact.

Great Expectations has been the subject of numerous critical studies; its treatment of parenting has generated a particularly rich strain of literary analysis.[166] What I propose here is just a footnote to these studies, a modest (or not so modest) reminder of how firmly, if imperceptibly, our readings of this novel are grounded in our cognitive biases. Bering observes that if "in its efficiency, [an adaptive cognitive] system has grafted itself onto the standard cognitive lens of the species, then it may actually be so transparent that it requires a great effort to see."[167] The rhetorical and emotional impact of *Great Expectations* is so bound up with our evolved cognitive adaptations for processing living beings differently from artifacts that it indeed requires a special effort to recognize these adaptations in action.

The story's central image of the person-as-an-artifact emerges from the relationship between Miss Havisham and her beautiful adopted daughter, Estella. Miss Havisham brings up Estella to "have no heart" and "to wreak revenge on all the male sex" by breaking *their* hearts—by having them fall in love with Estella and see her toy with and despise their tender feelings.[168] Molding Estella from her early childhood into an instrument of her will is Miss Havisham's way to get even with a cruel world, in which her own heart got broken when she was a young woman and a man whom she passionately loved betrayed and abandoned her.

Like the robot-surgeon in Asimov's "Bicentennial Man," Estella seems to be content with the "artifactual" view of herself. On one occasion, she tells Pip, "We have no choice, you and I, but to obey our instructions. We are not free to follow our own devices, you and I." On another, she calmly explains to her adoptive mother that she is constitutionally unable to love her in the traditionally sentimental way in which Miss Havisham wants to be loved by her child: "I must be taken as I have been made."[169] Estella's striking pronouncements feel intuitively plausible to us because they tap our tendency to associate "made" entities with rigidly defined functions and a severely circumscribed range of emotions. To put it at its starkest, the same cognitive adaptations that tell you that the cut in two thrickler will *not* die or feel anything are responsible for your intuitive understanding that, because Estella is an instrument "made" by Miss Havisham, her emotional capacities may indeed be limited in comparison to those of other people.

Or they may not. Estella may not know herself well (as Pip fervently hopes she does not). Young as she is, she may not be aware of recuperative powers that could expand her emotional potential in spite of her bizarre upbringing. After all, Estella is not a robot, and we are not in the realm of science fiction. What is important, however, is that Estella's current statements and beliefs about her irredeemable heartlessness *make sense* to us. And they would not have made any sense had we not had a suite of cognitive adaptations that strongly encourage us to associate made objects with limited functions oriented toward fulfilling the intentions of their maker. (A person who makes a chair intends it for sitting, and that's what the chair is for—no more and no less.)

Moreover, let us see how functionalism colludes with essentialism in *Great Expectations*—for the two are never neatly separated in literary works. Estella's conviction that, though physiologically speaking she has a heart, psychologically speaking, she has none—for her mother has "put ice in its place"—appeals to the same cognitive adaptations of readers as does Marge Piercy's carefully casual observation that Avram has an "artificial heart."[170] Because the heart figures so prominently in our conceptualization of individual essence, hearing that a person has "no heart" or a piece of "ice in its place" makes it easier for us to make sense of the person's "essential" psychological difference. Estella is different because her heart has been "stolen."[171] As such, she can indeed be conceptualized on some level as an artifact—an effective tool for breaking people's hearts—and it is the final irony of the novel that one of the hearts she breaks without ever intending it (for, remember, tools have no theory of mind and thus no notion of harming other people!) is the heart of her adopted mother.

(Note, too, that in Dickens's revised happy ending to *Great Expectations*, when Pip and Estella meet again and Pip hopes that they will not part anymore, Estella refers to her heart in very different terms. Not "stolen" anymore, not "a piece of ice," it is now open to feelings and memories. As she puts it, "I have given a [remembrance of you] a place in my heart." Estella still speaks of herself in somewhat functionalist terms, but these terms are framed—and overridden—by the references to hearts and emotions: "[My] suffering has been stronger than all other teaching, and has taught me to understand what your heart used to be. I have been bent and broken, but—I hope—into a better shape. Be as considerate and good to me as you were, and tell me we are friends."[172] The revised ending thus suggests a possibility of an escape from functionalism for Estella, a possibility stronger than that of the original ending, in which Pip speaks hopefully of Estella's heart, but she herself does not.)

Hilary M. Schor observes that "the language of 'making' is pervasive" in *Great Expectations,* and it "is no less present" in the "interwoven stories" of Pip and Magwitch: "the question of who's to be maker is as powerful as who's to be master."[173] Magwitch is yet another surrogate parent of the story, who strives to mould his adopted child into something that he himself is not and cannot ever be. Magwitch works hard in Australia to pay for the gentlemanly education of his young protégé of working-class origins back in England. As he explains to Pip when they finally meet, at the hardest moments of his life it was an emotional "recompense" to him "to know in secret that [he] was making a gentleman." Magwitch is ecstatic when he beholds the adult Pip—"this is the gentleman what I made! The real genuine One!"[174] The ex-convict's happiness and sense of self-fulfillment thus hinge exclusively on the object of his making—while that object can hardly conceal his horror and revulsion at the sight of his creator. As Pip puts it, evoking Shelley's *Frankenstein* and reversing the relationship between Victor Frankenstein and his monstrous progeny, "The imaginary student pursued by the misshapen creature he had impiously made, was not more wretched than I, pursued by the creature who had made me, and recoiling from him with a stronger repulsion, the more he admired me and the fonder he was of me."[175]

Pip's initial negative response to Magwitch and his attempt to reverse the dynamics of power between the maker and its creature are clearly influenced by his class snobbery. As a "gentleman," particularly of such recent coinage, Pip is terrified of being associated with a lower-class criminal. However, something else is happening here, too. Pip's refusal to accept his role as Magwitch's "creature" represents an important contrast to Estella's relatively placid acceptance of her position as a "made" entity. Generally, throughout the story, Pip seems to manifest more resistance than Estella does to being treated as an artifact. For example, he actively resents the fact that Miss Havisham used him as "a model with a mechanical heart to practise on when no other practice was at hand"—Estella, who has been used by Miss Havisham in the same fashion, never voices such complaints.[176] Similarly, in his delirious dreams Pip is shown to struggle against being taken for a mere object. As he puts it, trying to describe those strange visions,

> [I] confounded impossible existences with my own identity[;] . . . I was a brick in the house-wall, and yet entreating to be released from the giddy place where the builders had set me; . . . I was a steel beam of a vast engine, clashing and turning over the gulf, and yet . . . I implored in my own person to have the engine stopped, and my part in it hammered off.[177]

Finally, in one of his desperate attempts to fight objectification, he tries to turn the tables and use the rhetoric of functionalism against those who would objectify him. Here is Pip implying that by treating children in her power as artifacts as a consequence of her wounded pride and delusion, Miss Havisham herself has violated the will of the Being who "made" her:

> That she had done a grievous thing in taking an impressionable child to mould her into the form that her wild resentment, spurned affection, and wounded pride, found vengeance in, I knew full well. But that in shutting out the light of day, she had shut out infinitely more; that, in seclusion, she had secluded herself from a thousand natural and healing influences; that her mind, brooding solitary, had grown diseased, as all minds do and must and will that reverse the appointed order of their Maker; I knew equally well.[178]

The passage is centrally concerned with Miss Havisham's distorted emotions, but note the tacit appeal to functionalism in the last lines. Pip cannot claim directly that Miss Havisham was made by God with a certain function in mind. Nevertheless, by invoking the "appointed order" of her "Maker," Pip implies that Miss Havisham *ought to* have curbed her individual will and brought it in line with the Maker's general vision of his creatures as open to the "natural and healing influences" of life. That she did not curb it and has thus violated the "function of living," so to speak, is her crime against her Maker.

Several critics have argued that Estella also resists being objectified by her adopted mother, by her future husband, and by Pip himself.[179] Still, compelling as their arguments are, it seems that the author presents Pip's resistance as more vocal and immediate than that of Estella. We can read it as Dickens's acquiescence to the tradition of portraying women as generally more submissive than men. (For as Rousseau would have it, "Woman is made to yield to man and to endure even his injustice. You will never reduce young boys to the same point. The inner sentiment in them rises and revolts against injustice. Nature did not constitute them to tolerate it.")[180] Or we can turn to other factors that played a role in Estella's and Pip's upbringing. For as Phelan observes, "Pip had the advantage of exposure to Joe Gargery and that seems to be as significant to the differences between Pip and Estella as their gender differences."[181]

But whether we think that Dickens upholds this patriarchal creed or, on the contrary, remind ourselves that his other novels feature women who refuse to tolerate injustice and personal subjugation, from a cognitive angle, one thing seems clear: the weight of functionalist rhetoric bears much more on Estella than it does on Pip. The novel's intuitive appeal to our evolved cognition forces us to see as logical Estella's unhappy compliance with her rigidly defined role just as it encourages us to sympathize with Pip's struggle against objectification. After all, being made into a *gentleman* (a category that we tend to essentialize) is rather different from being made into a tool for heart breaking. The former opens up possibilities—*for what is the "function" of a gentleman?*—and the latter forecloses them: if your "function" is to break hearts, then hearts you break. The difference in the rhetoric of functionalism when applied to Pip and Estella thus seems to implicitly naturalize both Pip's resistance and Estella's compliance.

And yet there is more to it, for *Great Expectations* is not simply an exercise in rhetoric but also an immensely complex cultural artifact reflecting and shaping Victorian social class awareness. Being made into a gentleman presupposes a lucky escape from the particular kind of functionalism implied by an inferior class standing. As a gentleman, Pip won't have to work; his blacksmith days are safely behind him, and he will lead the life of leisure and exercise his free will daily. But viewed from this perspective, Miss Havisham's treatment of Estella may *also* seem liberating rather than confining. Being a tool of Miss Havisham's deranged will and plying the trade of heart breaking in sumptuous ballrooms and sitting rooms might still be well preferable to the life of backbreaking physical labor and severely limited choices available to a serving-class woman in nineteenth-century England. Our awareness of Estella's working-class origins renders difficult any unambiguous pronouncement about the meaning of her apparent compliance with and/or resistance to the will of her mad maker. After all, her surrogate mother did snatch her from the fate of *industrial tool,* thus making her into *a person* with a much wider range of options.

Or did she? The "cognitive" collides with the "social" in *Great Expectations* in a way that renders this question nearly impossible to answer. Because on some crucial level we process information about natural kinds and artifacts through different cognitive domains, we are intuitively jolted when these two domains are momentarily brought together and a human being is characterized as "made" in the fashion of an artifact. By forcing us to see a person as an object with a rigidly defined function, writers thus open up new interpretive spaces, encouraging us to grapple forever with questions that we would not have even considered otherwise. Is Estella freed from social objectification ("working-class women are made to work") only to be plunged into personal objectification ("Estella is made for wreaking revenge on the male sex")? Is she liberated from one kind of gender objectification ("Women are made to please men") only to be weighed down with another ("Estella is made to destroy men")? Is this labyrinth of functionalism escapable, or are we to wander in it forever, goaded by the adventuresomeness of our cognitive architecture, which always encourages us to see what happens if we cross the domains of living beings and artifacts in this way or that way?

PART THREE

Some Species of Nonsense

1. *How Nonsense Makes Sense in* The Hunting of the Snark

Nonsense makes sense. From the cognitive perspective advocated by this study, this paradoxical statement has more truth to it than we may have ever imagined. Consider the "types of impossibilia," listed by Noel Malcolm in his book *The Origins of English Nonsense* (1997), which were already present in medieval nonsense poems: "The reversal of roles by animals (hens seizing hawks), the animation—or, to be more precise, the animal-ization—of inanimate objects (flying millstones being particularly common), and the performance by animals of complex human activities (such as spinning or building)."[1]

You can see how all three types build on our evolved cognitive adaptations for categorization. The reversal of animal roles plays with our essentialist proclivities: because it is "in the nature" of hawks to prey on hens and it is "in the nature" of hens to be wary of hawks, a turnaround in these "natural" hierarchies is bound to strike us as fascinating. Similarly, the performance by animals of complex human activities crosses the subcategories of "animal" and "human" nested within our larger category of "living beings." Because it is not "in the nature" of animals to engage in activities that we associate with human beings, the image of such an animal remains perennially attention-catching. Finally, the "animal-ization of inanimate objects" presents us with entities that cannot be fully assimilated either by the cognitive domain of artifacts or by the domain of living beings.

Such animated artifacts are of course closely related to the robots and cyborgs of science fiction as well as to various talking objects in cartoons. The crucial difference between the two is that science fiction stories and cartoons typically provide us with descriptions of the imagined worlds in which such counterontological entities might exist. Or, to put it differently, they provide us with enough textual cues to engage in what Jonathan Culler has called "naturalization" on the first level of verisimilitude.[2] That is, when we approach a fictional narrative, our default expectation is that the world that it depicts will "derive directly from the structure of the world."[3] Should the narrative then "explicitly violate" our expectations and present us with entities and events impossible in the "real world," we start naturalizing it. In other words, "we are forced to place the action in another and fantastic world," thus constructing an alternative reality whose rules will now constitute a verisimilitude of its own.[4]

Note that this still does not help us to process any of our counterontological entities within a single conceptual domain, but at least we have bracketed their "reality" as systematically different from ours, on several counts. We thus "know" that in Norika (in Piercy's *He, She and It*), the technology has advanced to such a degree as to make possible embodied artificial intelligence, that the dancing candelabra, mantel clock, and teapot are in fact people transformed into objects by the enchantress in Disney's *Beauty and the Beast,* that the talking rabbits, mice, birds, and caterpillars inhabit the special world of "Wonderland," the whole of which later turns out to be Alice's dream, anyway (a *double* insulation, so to speak, of our world from the world swarming with ontological violations), and that because we are watching an animated feature or reading a fairy tale, we should be prepared for such chimeras. By contrast, nonsense poetry offers us little or no such framing. It thrusts the "flying millstones" in our faces and blithely challenges us to deal with their strangeness as well as we can.[5]

Hence we learn from "the earliest known nonsense . . . written by a German Minnesinger, 'Reinmar der alte,' who died in 1210," that "[b]reastplate and crown want to be volunteer soldiers."[6] Try as we may (and we have been trying for eight hundred years now) these ambitious artifacts will never be fully assimilated in our minds as your regular everyday breastplate and crown. These two specimens possess a humanlike theory of mind. If they are capable of *wanting* to be volunteer soldiers, it means (our essentialism kicks in) that they are also capable of feeling tired, jealous, curious, flabbergasted, socially inferior or superior, thankful for being cheered up on a bad day, resentful for being thought stupider than they perceive themselves to be, and so forth. No human mental state is alien to them.

Then there is also the "'Lügendichtung' genre of German nonsense poetry" (I am quoting from Malcolm again) that "enjoyed a long life" and reached "its high point" in the fourteenth and fifteenth centuries. One "very popular example from the fourteenth century was the 'Wachtelmäre'"—a "long and elaborate narrative, [which] tells the story of a vinegar-jug who rides out to joust against the

King of Nindertda in the land of Nummerdummernamen, which lies beyond Monday."[7] I turn to this example to illustrate, among other things, what I mean by the title of this part—"*Some* Species of Nonsense" (my emphasis). Although I do think that *any* successful piece of nonsense engages our cognitive predilections, I focus here on only a few types of nonsense. I do not analyze, for example, the nonsensical images that conflate time and space (e.g., "the land which lies beyond Monday") or those that build on a word play (e.g., "Nummerdummernamen"). The vinegar jug who rides out to joust against a king, however, is just my jug of vinegar—an artifact that reaches over to the domain of human beings and gets stuck in between the two domains—a unique, wishing, feeling, thinking, if rather too-pugnacious-for-its-own-good artifact.

But if you want to see an entity stuck not just between *two* domains, but also among *three* domains, and as such rebuffing with a particular ease all our attempts at assimilating it within any of those domains, look at the title "protagonist" of Lewis Carroll's classical nonsense poem *The Hunting of the Snark* (1876). Here are the opening stanzas:

> "Just the place for a Snark!" the Bellman cried,
> As he landed his crew with care;
> Supporting each man on the top of the tide
> By a finger entwined in his hair.
> "Just the place for a Snark!" I have said it twice:
> That alone should encourage the crew.
> Just the place for a Snark! I have said it thrice:
> What I tell you three times is true.
> The crew was complete: it included a Boots—
> A maker of Bonnets and Hoods—
> A Barrister, brought to arrange their disputes—
> And a Broker, to value their goods.[8]

This opening contains an unfamiliar concept—"Snark." To understand how the introduction of such a concept affects us, think again of the fake words from our discussion in part 2: zygoons, thricklers, and kerpa. When we hear that 1) "zygoons are the only predators of hyenas," 2) that "thricklers are expensive, but cabinetmakers need them to work wood," and 3) that to make a certain dish, we have to take one pound of boiled spinach, two teaspoons of butter, a pinch of salt, and a spoon and a half of kerpa, we are immediately able to deduce that 1) a zygoon will most probably die if you cut it in two, 2) that a cut-in-two thrickler might be of use to a thrifty cabinetmaker, and 3) that it might be possible to transport kerpa on a plane in a plastic jar.[9] Although we have never heard these words before (indeed, they do not exist), our cognitive adaptations allow us to categorize a zygoon as a living being, a thrickler as an artifact, and kerpa as a substance, and by doing so to gain immediate access to a large set of inferences about each of these entities.

Now "Snark" is also a word that does not exist, which means that on first hearing it, we are on the lookout for information that would allow us to place it within a familiar category and immediately start inferring new things about it. Carroll is intuitively aware of this, and he plays with us by seemingly telling us something about the Snark but really giving us no inference-building material at all. By the beginning of the third stanza, we have heard the word "Snark" mentioned three times, accompanied by the assurance from the Bellman that what he tells us "three times is true," but we have no leads for inferences!

Well, *almost* no leads (and now we are scrambling desperately), for apparently "Snark" is valuable enough to get together to hunt for. Although even this is not at all obvious—the Snark-hunting "crew" that Carroll then proceeds to describe is such a odd crew that what *they* may consider worthy of obtaining might not be considered so by others. This brave band of bachelors includes the Bellman, the maker of Bonnets, the Barrister, the Broker, the Billiard-Maker, the Banker, the Beaver, the Baker, and the Butcher. Many of them are incompetent (the Bellman buys a blank map; the Baker can't really bake; the Butcher can only butcher Beavers), and as we learn more about them, it becomes nearly impossible to speculate about the qualities of the "Snark" that brought them all together.

But, finally about sixteen pages later (and I think one reason that Carroll can afford so long an interruption is that he intuitively knows that we've been hooked and are now looking and looking for any inference-building cues), we get back to the Snark proper. For the benefit of his crew members, the Bellman lists the five essential qualities of the "warranted genuine" Snark:

> Let us take them in order. The first is the taste,
> Which is meager and hollow, but crisp:
> Like a coat that is rather too tight in the waist,
> With a flavour of Will-o-the-Wisp.
> Its habit of getting up late you'll agree
> That it carries too far, when I say
> That it frequently breakfasts at five-o'clock tea,
> And dines on the following day.
> The third is its slowness in taking a jest.

> Should you happen to venture on one,
> It will sigh like a thing that is deeply distressed:
> And it always looks great at a pun.
> The fourth is its fondness for bathing-machines,
> Which it constantly carries about,
> And believes that they add to the beauty of scenes—
> A sentiment open to doubt.
> The fifth is ambition. It next will be right
> To describe each particular batch:
> Distinguishing those that have feathers, and bite
> From those that have whiskers, and scratch.

Later on in the poem we get another relevant description of the Snark, this time provided by the Baker:

> I engage with the Snark—every night after dark—
> In a dreamy delirious fight:
> I serve it with greens in those shadowy scenes,
> And I use for striking a light.[10]

Here is, then, what I mean when I say that good nonsense makes complete sense from a cognitive point of view. Hearing of something referred to as "it" and learning that it is "crisp" and "hollow" to the taste, we assume, as we would with kerpa, that the Snark is a substance, most likely a foodstuff. That allows us to make a series of inferences about the Snark that we usually make about edible substances (e.g., some people may find it tastier than others; it may spoil with time; it can be cooked, though it may also be consumed raw, etc.). Immediately after, however, we are bombarded with information about the Snark's habits and personality traits: laziness, bad sense of humor, love of unwieldy gadgets, ambition, and, later yet, pugnacity. This leads us to assume that the Snark is a creature with a fully developed theory of mind—capable of an infinite array of human mental states.

Generally, once we have assumed that about *any* entity, no subsequent information can force us to classify that entity with mere birds, animals, substances, and artifacts, even though we may learn that some Snarks "have feathers, and bite"; that some "have whiskers, and scratch"; that some are "serve[d] with greens"; and that some can be used "for striking a light." We are thus not surprised to hear that one of the members of the team (fittingly, the Barrister) had a dream in which the Snark, "with a glass in its eye, / Dressed in gown, bands, and wig, was defending a pig / On the charge of deserting its sty."[11] For giving long speeches before a jury of one's peers is just one of innumerable activities that creatures with theory of mind—such as people—are capable of.

And yet through all this speech giving and pig defending, the Snark is referred to as "it," and we cannot simply forget that it tastes crisp, can be served with greens, and can be used for striking a light. The Snark thus remains an entity that crosses three conceptual domains—that of human beings, that of substances, and that of artifacts—and as such cannot ever be fully assimilated to any of them.[12] This accounts for the fact that the Snark remains subject to endless symbolic interpretations—by intuitively spreading its ontology all over our conceptual map, Carroll has ensured that his readers will forever be attempting to put this impossible creature "together again."

Let me now quote at length from one such attempt, that of Edward Guiliano, a prominent Carroll scholar. Guiliano reads the poem in the context of Charles Dodgson's overweening awareness of death (*The Hunting,* after all, was conceived when Dodgson was nursing his dying twenty-two-year-old cousin) and his general fascination with "states of being (which include not only dying but dreaming and even spiritualism)."[13] Hence Guiliano's interpretation of the Baker's dream:

> The Baker also describes what can be viewed as a state of existential dread:
>
> I engage with the Snark—every day after dark—
> In a dreamy delirious fight:
> I serve it with greens in those shadowy scenes,
> And I use it for striking a light.
>
> This stanza provides an excellent illustration of the tension that exists between the comic tone and the underlying terror that characterizes the poem for readers today. It also provides us with a glimpse at the nature of Carroll's artistic temperament. The first two lines, "I engage with the Snark—every night after dark— / In a dreamy delirious fight" can be read as a statement that parallels Dodgson's own experience. He was an insomniac kept awake at least partially by haunting and troubling thoughts. In fact, he published a book of puzzles devised to ease the pain of his sleeplessness—some mental work to help free his mind of its troubling thoughts, thoughts that surely occurred to many of his contemporaries as well. . . .
>
> But although we can find suggestions of Dodgson's anxieties in the Baker's dread, in the closing lines, "I serve it with greens in those shadowy scenes, / And I use it for striking a light," we find only whimsy and nonsense. They change the tone completely.[14]

Let me start by quarreling with Guiliano about one minor point before agreeing with his larger argument. While he views the closing lines ("I serve it with greens . . .") as "only whimsy and nonsense," I insist that they are profoundly logical and sensible, indeed *necessary* at this point in the poem. Considering Carroll's larger project—which I am now recasting in explicitly "cognitive" terms—he had to prevent his readers from getting too comfortable with the idea of the Snark as either a person, or as an artifact, or as a substance. The stanza in which the Baker first "engages" with the Snark every night after dark, then serves it with greens, and then uses it for striking a light hits roughly all three domains and as such forcefully reminds us that the Snark is not quite a person, not quite an edible substance, and not quite an artifact, even if it apparently has some qualities of each of the three.

I say that it hits *roughly* all three domains because of the ambiguity of the word "engage." "Engage" is not the strongest interactive verb out there—one can engage with an entity that does not have a theory of mind. "I argue with the Snark," for example, would imply the Snark's "personhood" stronger, but of course it does not scan. Moreover, Carroll can afford such relatively abstract verbs as "engage" in his reference to the Snark because by this point he has already firmly established in his reader's conscience that the Snark has a human-like theory of mind.

I have pointed out already that once we have assumed that a given entity has the potential for a broad variety of states of mind, it is pretty much impossible to make us classify that entity with something that lacks such potential. (And that is not surprising. Throughout our evolutionary history, it must have made more survival sense to overattribute capacity for intentionality to an entity that has shown some agency than to underattribute it. It's better to mistake a tree for an enemy than to mistake an enemy for a tree.)[15] Thus the word "engage" in the first line of the stanza implies that the Snark has some intentionality at least and specifically human intentionality at most, and it is this implication of a possibly human intentionality that is then undercut neatly when we are told that the Snark can be served with greens and can be used for striking a light.

To repeat, where Guiliano sees mere "whimsy and nonsense" (i.e., the intrusion of first the edible Snark and then the match-like Snark), I see a cognitive logic. But then it is precisely because I disagree with Guiliano on this point that I can endorse his larger view of the Baker as expressing his own, our, and Carroll's "existential dread." Because the Snark can never be fully assimilated to any single conceptual domain and yet at the same time consistently activates inferences from all three domains, our search for the meaning of the poem never stops. This search may lead us to inquire into the circumstances attending the writing of the poem, into the sociohistorical, cultural, and aesthetic milieus into which the Snark was "born," and into the state of mind of the person who wrote it and the person who reads it. It also may—as in this particular case—help us to understand and articulate our feeling of "existential dread." Moreover that feeling must be *intensified* rather than *dissipated* by the images of the Snark served with greens and used for striking a light. For it seems to me that the cognitive vertigo induced in us by the entity that resists—and continues resisting with every new description—conceptual assimilation should make even stronger that sense of helplessness and "existential agony" that several critics see as pervading the poem.[16]

I need now to address the limits of my "cognitive" analysis of Carroll's poem. For it may appear that any author can follow the recipe outlined above and write a brilliant piece of nonsense. Just present the reader with an entity that straddles several conceptual domains—two domains is good, three is apparently even better—and presto: the poem will hold our attention, will resist a settled interpretation, and will be beloved by generations of readers as Carroll's *The Hunting* has been.

You can see already the problems with this recipe. A rhymed narrative about a counterontological entity may indeed hold our attention by resisting a settled interpretation and thus qualify, technically speaking, as a nonsense poem and still remain a pedestrian exercise in rhyming. In fact, I can think of at least one contemporary author of children's nonsense poems who does follow this recipe very faithfully (intuitively, of course, not consciously: he began writing long before cognitive scientists came along with their research on essentialism and functionalism), but his work seems utterly lacking in charm, and it doesn't come close to approaching the magic of Carroll's *The Hunting.*

To some extent, this is a question of personal preference. Still, my larger point remains. There are many ways to tell a story about counterontological creatures, and the way Carroll does it is unique. I hope to have uncovered one underlying technique that he uses to keep the Snark mysterious; what I do not address here—and what is absolutely crucial for our understanding of the poem—is how this technique is intertwined with other aspects of *The Hunting.*

It matters, for example, that the poem builds a series of complicated relationships among its various alliterative bachelors; that the protagonists are concerned about their social status within the group; that the hunting adventure touches something deep in each of them and transforms some of them. We usually reserve such terms of analysis for novels, but they seem to be relevant in the case of *The Hunting*. Thus, a more comprehensive conversation about this specific poem (as opposed to just any nonsense poem) would inquire into the ways in which the emotional and social concerns of its protagonists interact with the ontological quandary implied by the image of the Snark. Or, to put it differently, from whatever interpretive angle you approach *The Hunting*, you may eventually want to ask how your interpretation engages with the fact that the central image of the poem, the nexus of the dreams, anxieties, and ambitions of its protagonists, is a cognitive conundrum that cannot be solved.

2. "Strings of Impossibilia" and What They Tell Us about the Value of Nonsense

Remember the belligerent vinegar jug that rode out to joust against the King of Nindertda in the land of Nummerdummernamen? It turns out that this poem is quite unusual for the medieval nonsense tradition. As Malcolm points out, it has "a unified narrative structure," while "most of the Lügendichtungen are little more than . . . strings of impossibilia, with images which are built up over a few lines at most."[17] Here is a more typical example from a late fifteenth-century Lügendichtung:

> A Swiss lance and a halbard
> Were dancing in a hop-field;
> A stork's leg and a hare's foot
> Were playing sweet dance-tunes on the pipe.[18]

Armed with our cognitive framework, we can see right away why these images strike us as nonsensical. The "lance and the halbard" (that is, halberd) are depicted as having intentionality, which makes them into conceptual hybrids straddling the domain of artifacts and the domain of living beings. Note, though, that the "stork's leg" and the "hare's foot," of which we hear next, are not exactly artifacts. That is, we can easily approach them from a functionalist perspective when we consider their relationship with the bodies that they serve (as in "legs and feet are made for locomotion"), but still they are not man-made objects on the order of lances and halberds.

The animated objects transcending their original function are thus paired here with the animated body parts transcending theirs, and the overall effect of such a pairing is the disruption of the opening narrative about the Swiss lance and a halberd. We may insist on reading some narrative coherence into the stanza by observing that both pairs of hybrids engage in some form of merrymaking, but still the tone established by this string of impossibilia is such that we cannot predict what new chimeras and what straddling of conceptual categories will come next. It could be anything.

Strings of impossibilia were popular not just in the German literary tradition, but also in the French and English. (And I can attest to their presence in twentieth-century Russian nonsense poetry for children.) Malcolm sees the cross-cultural prevalence of such rhymed antinarratives as evidence of the ongoing exchange between different national traditions of nonsense poetry. As he observes,

> The very idea of putting together strings of impossibilia . . . simply for the effect of comic absurdity which they produced on their own . . . is fundamental to both the German and the French poems, and it is very unlikely that the idea was just invented independently on two occasions, within the same century, in two neighboring countries. Literary influences flowed to and fro between the French and German vernaculars throughout the Middle Ages.[19]

Coming from a cognitive perspective, I propose a theory that complements Malcolm's explanation of the cross-national popularity of strings of impossibilia. Look again at the description of the Swiss lance and halberd. When we learn that these two weapons are "dancing" in the "hop-field," we may take it not as a nonsensical statement but as a metonymic description of a raging battle, something along the lines of the Homeric "Nine days the arrows of god swept through the army."[20] This *commonsensical* interpretation becomes impossible, however, once we hear of the stork's leg and hare's foot playing dance tunes on the pipe. The addition of a differently structured nonsensical image—that of the animated body parts—ensures that each unit of the poem is processed as impossibilia.

Let us see how it is done in a different string, this one coming from the English literary tradition. Here is a poem called "A Fancy" from the London miscellany *Sportive Wit: The Muses Merriment* (1656):

> When Py-cryst first began to reign,
> Cheese-pairings went to warre,
> Red Herrings look't both blew and wan,
> Green Leeks and Puddings jarred.
> Blind Huh went out to see
> Two Cripples run a race,
> The Ox fought with the Humble Bee,
> And claw'd him by the face.[21]

Again, taken by themselves the first two lines, "When Py-cryst first began to reign, / Cheese-pairings went to warre," could be interpreted as a political allegory by a discerning reader eager to see in them some reference to the events of the English Civil War of 1642–51. To squash the ambitions of that discerning reader, the anonymous author(s) of the poem disrupted its narrative unity by adding other differently structured units of nonsense. Interpreted individually, each of those units can be perceived as having some topical meaning; their accumulation, however, forecloses such an interpretation.[22]

In other words, I agree with Malcolm's argument that literary influences played a crucial role in making the string of impossibilia a cross-cultural phenomenon. At the same time, I think that the reason that the authors of nonsense poetry were so willing to borrow this particular gimmick from their foreign colleagues was that it helped them to preserve the nonsensical thrust of their verse.

For if a nonsense poem hopes to *stay* a nonsense poem it has to fight its readers all the way. When confronted with its incongruities, we immediately start *interpreting* it—that is, casting around for ways in which it could be incorporated into our familiar conceptual frameworks. To quote Culler again, "In poetry deviations from the [verisimilar] are easily recuperated as metaphors which should be translated or as moments of a visionary or prophetic stance."[23] Granted, we cannot assimilate the images from "A Fancy" within the usual ontological categories—a piecrust that rules the country cannot be filed away as either a foodstuff or a person—but we can view the poem as a political allegory or a symbolic comment on the human condition and so take its nonsensical edge off. (We have already seen how this can be done, with *The Hunting of the Snark*.)

But stringing along bits of unconnected or loosely connected impossibilia effectively shuts the enterprising interpreter up. Deviations refuse to be recuperated as translatable metaphors; the posture of prophetic stance is better not attempted when the "visions" feature piecrusts, cheese pairings, and green leeks.

This, of course, reopens the question of some intrinsic "value" of nonsense. For why go to such lengths to keep the reader stuck with the apparently inassimilable images? Why sacrifice such an important aspect of the verse as its narrative unity (which, after all, helps its transmission and survival) to keep nonsense nonsense?

To begin to answer this question, consider this: I have focused so far on fictional narratives built around various "strange concepts." But strange concepts make possible all kinds of narratives, both fictional and nonfictional. We are constantly dealing with strange concepts, which range from products of our casual everyday objectification and anthropomorphizing to elaborate chimeras populating our religious representations, cartoons and other visual arts, science fiction, and nonsense poetry. We try to assimilate them within our familiar categories, sometimes failing (as in the case of the Snark), sometimes succeeding to a degree, which may make us change our thinking for better or for worse (e.g., hearing that "women are made to bear" may prompt one to question her views on matrimony).[24]

This leads me to suggest our strange concepts are not merely superfluous fancies, the imaginative surplus that we can afford in our idle moments but could as well do without. On the contrary, I believe that both the making and the interpretation of such concepts are crucial to our cognitive well-being. For our cognitive architecture often carves the world at joints that are not there, for instance, insisting on "essences" that separate one species from the other and on the rigid distinction between animated and nonanimated entities.[25] Contemplating concepts that challenge our cognitive biases may help the mind to retain its flexibility and its capacity for responding to its infinitely complex and changing environment.[26]

So perhaps we need our strange concepts and may sometimes work quite hard on preserving their strangeness for as long as we can. To take this argument further, I would say not only do we need strange concepts but also that we may benefit from a smaller subclass of *really strange concepts* that resist assimilation all the way, even as their constituent elements retain their grounding in our familiar ontologies. (That is, a Swiss lance still activates in us inferences associated with artifacts, and the act of going dancing still activates in us inferences associated with living beings, which is precisely the reason that the juxtaposition of the two is jarring.)

One possible reason that it might be good for us to be confronted now and then with *really strange concepts* is that they force us to be aware of our own processes of categorization. For surrounded as we are with strange concepts in our everyday life, we often process them so swiftly that their salubrious jolting effect on our cognition might be diminished. Moreover, we get used to many of them as we do to familiar metaphors. When in the early 1930s Joseph Stalin first called socialist realist writers "engineers of human souls" (and Maxim Gorky happily echoed it with his own concoction: "craftsmen of culture"), it must have been jolting, and now, with the benefit of a cognitive evolutionary perspective, we know why. You can "engineer" an artifact; you don't "engineer" a person's "soul." If you do, either you are a very special kind of engineer—what Stalin's propaganda machine presented as a unique Soviet writer—or the souls that you make are very special kinds of souls: unique Soviet souls! But startling as the expressions "engineers of human souls" and "craftsmen of culture" might have been initially, that effect largely wore off over the years, especially since Soviet culture was saturated with images of machinelike people: tough but easily replaceable. By the time I was in high school, I heard and used both phrases without giving them much thought or attention.

But with the *really strange concepts* you just can't do that! They demand your attention every time you come across them, and by doing so make you more aware of your cognitive processes. Cognitive psychologist Peter Carruthers and cognitive literary critic Reuven Tsur have written separately on the importance of such awareness.[27] Tsur in particular has argued that fictional narratives affect us by delaying or disrupting "in some other manner" our cognitive processes, and that this act of disruption may derive some of its pleasurable effect from confirming that our cognitive adaptations are functioning well.[28]

Applying Tsur's insight to our present discussion, I suggest that really strange concepts, such as those contained in strings of impossibilia, force us to pause amid our everyday unconscious mental processing and look about ourselves, so to speak. When we are faced with the halberd dancing in a field and cannot file it away as some kind of a metaphor or allegory, we know that something strange is going on, and, moreover, we know that we know that something strange is going on. (What readers do by way of explaining those strange images is say that they represent "the world upside down"—not the most meaningful explanation when you think about it.) The realization that *we know that we know* serves as indirect evidence that our categorization processes are functioning well. They do not let through an incongruous image—they stop that image at the gate of our unconscious.

Two clarifications are in order. First, by saying that we thus get evidence that our categorization processes are functioning well, I am not saying that this means that they reflect the world accurately. I have already observed that they do not—they saddle us, for example, with such false intuitions as the perception of "essence." A better way to describe their well-functioning state is to say that they reflect the world not accurately but *adequately*—they are far from ideal, but they are "good enough" to get us through the day without a major mishap.[29] The image of a dancing halberd thus assures us that, imperfect as they may be, our cognitive adaptations for categorization are in gear.

Now to the second clarification. When I say that the value of strings of impossibilia (and hence the cross-cultural popularity of the form) resides with their ability to protect and preserve really strange concepts, I do not mean that there is something culturally predetermined about this genre. Neither do I imply that the cultures in which this literary form does not exist are somehow less "cognitively astute" than those in which it thrives. In *Why We Read Fiction,* I discuss at length the issue of the historical "inevitability" of genres that engage our cognitive capacities in particularly focused ways, so let me quote from that study, substituting the cognitive capacities in question for the theory of mind (the subject of inquiry in *Why We Read Fiction*) :

> There is nothing really ensured or determined about how genres arise, metamorphose into other genres, or die out, even if they do "get at" our [categorization biases] in a particularly felicitous way. For all that we know, there might have been a man or a woman in [our past] who wrote an experimental [work of fiction] that could have started a new literary tradition [engaging our categorization processes] in a wonderfully unpredictable fashion. That [work] did not find an audience; or was lost in the mail; or its author changed his / her mind and never revisited this particular style . . . in his / her subsequent [writings]. Literary history reflects only a tiny subset of realized cognitive possibilities constrained by the myriad of local contingencies, and those contingencies include personal inclinations and histories of individual writers and readers.[30]

Moreover, other arts may offer different formats for preserving the shock value of strange concepts. The next section will consider one of those formats and conclude by attempting to make sense of a particular artwork. And fail, of course.

3. "Painters of the Unimaginable," or More about Really *Strange Concepts*

My argument is useful only as long as we remain clear about its limits. When I talked about nonsense poetry, I insisted that my analysis applied only to *some* nonsensical structures and, in the case of *The Hunting of the Snark,* only to one, very specific aspect of the poem. The same principle holds in my present cognitive inquiry into surrealist art, as it was practiced under the watchful eye of André Breton from the early 1920s to the 1960s. Surrealism came in so many different forms and personal styles, and featured so much evolution of personal styles, that I would not dream of trying to explain all of it via my cognitive framework. I am arguing instead that such a framework could elucidate the effect of *some* visual concepts that surrealist artists strove to express in their work and that as long as we are in agreement about the strictly defined scope of its application, it represents a useful interpretive tool.

For Breton, surrealism was a way of life—an ongoing revolution of perceptual practices. In particular, the aim of surrealist art was to "express the unconscious activity of the mind," resisting "as long as we live" its intelligibility along the comfortable lines of familiar everyday experience.[31] As Breton wrote, quoting pointedly a 1913 text by Giorgio De Chirico, a surrealist painter who by the late 1920s would arguably forfeit his claim to "true" surrealism because his art began to conform to "human proportions":

> For a work of art to be truly immortal, it must thoroughly transcend the limits of the human: common sense and logic must be lacking. In this manner it will be more like a dream, closer to the mentality of children. . . . The revelation we have of a work of art, the conception of a painting reproducing whatever it might be, which has no sense on its own, no subject, which from the point of view of human logic *doesn't mean a thing*—that revelation, that conception *must* be so strong within us that they procure such joy or such pain that we are obliged to paint, driven by a force greater than that which drives a famished person to throw himself like an animal onto the piece of bread he has suddenly found.[32]

Yet to "transcend the limits of the human" did not mean to abandon oneself to conceptual and interpretive nihilism (hence Breton's disaffection, in the early 1920s, with the Dada movement). On the contrary, as Breton put in 1934, surrealism had "always . . . expressed for [its true practitioners] a desire to deepen the foundations of the real, to bring about an even clearer and at the same time ever more passionate consciousness of the world perceived by the senses." To achieve these effects, surrealists attempted to juxtapose "interior reality" (the unconscious workings of the mind) with "exterior reality" (the result of conscious and habitual perception) and to confront "these two realities with one another on every possible occasion."[33]

Breton's explication of the surrealist project does not map neatly onto my cognitive theorizing. Hence my using a sentence from his *Nadja* (1928) as an epigraph for this book represents a bit of a sleight of hand. However, I am struck by how gracefully that sentence seems to express one of my main arguments, which is that our cognitive adaptations for categorization ("a mind's arrangement in regard to certain objects") may structure our emotional responses to certain aspects of our cultural representations (our "regard for certain arrangements of objects"), even while I do not claim that Breton was developing some protocognitive evolutionary stance. Note, too, how well the Freudian concept of the unconscious, which Breton took quite seriously even while he was aware of the limitations of psychoanalysis, meshes with our present understanding of cognitive processes as closed off to our conscious awareness yet powerfully impacting our perceptual and interpretive practices.[34]

In other words, there is a cognitive method to some surrealist madness, and it is because the artists intuitively followed that method that their work may claim to "deepen the foundations of the real." Consider, for example, Joan Miró's *The Carnival of Harlequin* (1924), which features, according to fellow surrealist Sarane Alexandrian, "an extraordinary fancy-dress ball, where not only human beings, but also animals and everyday objects, are wearing masks" (fig. 5).[35] When somebody's wearing a mask, it signals not merely a deliberate concealment of his or her facial expression but also a public "announcement" of the *intention* to conceal one's mental state (our faces being a crucial source of information about our thoughts and feelings). In its pointed intentionality, mask donning is thus distinct from mimicry as practiced by many animal species. It implies a fully developed human theory of mind on the part of the masquerader, and it is the presence of that intentionality that transforms the cat, the fish, and the several insects populating the painting into conceptual hybrids. For each of them is thus part animal and part human, and none can be fully assimilated to either of those subdomains within the larger domain of living beings.

Now turn to the ladder in the far left corner, the one with no rungs on the bottom and a big colorful ear sticking out from its upper part. The missing rungs are important: they speak to our functionalism, making us uncertain about the ontology of this erstwhile artifact. Is the ladder that you cannot use for climbing still a ladder or is it now some other object "masquerading" as a ladder? The curious ear complicates matters further. It implies either a sense of hearing that we associate with living beings, both animals and humans, or—if we choose to focus on how colorful the ear is—a fully developed theory of mind that we associate with humans. For if the ear is not merely an ear but a part of the ladder's "costume," it signals the ladder's intention to participate in the carnival and even "mask" itself as something else.

The color scheme of the painting also emphasizes the hybrid nature of the portrayed entities. Miró uses the same color (light slate blue in the reproduction of the painting that I have in front of me) to represent an artifact (the table on the right), one of the insect's wings, the body of the guitar-playing creature in the center, and the right half of Harlequin's face. Similarly, the coloration of the cat on the far right echoes that of the guitar. Finally, the ladder's ear is part red (the same red that covers the other half of Harlequin's face), part green (the same green as that of the levitating sphere on the right), part black (the black of the

Figure 5. Joan Miró, *The Carnival of Harlequin* (1924–25). Oil on canvas. 66 x 90.5 cm. Albright-Knox Gallery, Buffalo, NY.

long arm transecting the painting in the middle), and part yellow (the yellow of the body of a little merry mammal—or rodent?—on the bottom of the painting). The use of the same color to portray an animated being and an artifact deepens the impression of ontological uncertainty pervading *The Carnival.*

It is difficult to say how much of this uncertainty we register consciously when we simply take in the picture as a whole without trying to analyze it; say, when we pass by it in a museum.[36] A fleeting glance at *The Carnival* leaves one with an impression of overflowing activity but also with a peculiar feeling of being watched. For we *are* being watched by its numerous creatures (some of whom also listen quite attentively—as the big ear attests), who themselves are protected from our inquiring gaze by their masks. They see us—and know us—better than we do them, an impression that is further intensified if we pause by the painting and take a longer look and begin to realize (so to speak—for much of this realization remains unconscious) that we cannot safely "place" any of those entities in a familiar category.

Is that all there is to *The Carnival of Harlequin?* Of course not! I haven't even begun to discuss various symbolic meanings and cultural references implied by the interplay of shapes and colors of the painting. Nor do I intend to—for that has been done and will continue being done by professional art critics and historians, who are much better at it than I. I just want to point out that dissimilar as their interpretations may be, they will all draw on our differential conceptualization of living beings and artifacts. For what is our rhetoric of "objectification" and "anthropomorphization"—which is invariably present in the analysis of surrealist art—but an attempt to put into words our intuitive awareness of the patterns underlying our categorization processes? I see my contribution as bringing those patterns into the open and showing that the elements of surrealist artwork that we may view as nonsensical actually make a lot of sense from a cognitive perspective. They are, in fact, tightly bound by the regularities of our categorization processes.

Think too of the surrealists' abiding fascination with objects. As Salvador Dalí saw it, "the surrealist object serves no purpose other than to deceive, to extenuate and to cretinize mankind[;] . . . it exists only to honor thought."[37] Among such cretinizing or thought-honoring entities, surrealists counted the found object, the natural object, the interpreted found object, the interpreted natural object, the readymade, the incorporated object, the phantom object, the dreamt object, the box, the optical machine, the poem object, the mobile and mute object, the symbolically functioning object, the objectively offered object, the being object, and others.[38] Below I focus on only three of those categories—the readymade, the phantom object, and the natural object—because, as I have already observed, the theoretical framework that I am offering in this study is useful only as long as we remain selective about its application and do not foist it on everything that seems to resonate with it in some vague way.

According to Alexandrian, the term readymade "can be applied only to an industrially mass-produced object whose function is altered, and which is dragged from its context of automatic reproduction in the most ingenuous way possible." At this point, the best-known example of a readymade is perhaps Man Ray's *Gift* (1921), "a flat-iron with its ironing surface bristling with nails" (fig. 6).[39] If we "translate" the striking visual effect of such objects into our cognitive terminology, we can say that they tease our strong predisposition to think of artifacts in terms of their function. Once the function is subverted as radically as it is in the case of the flatiron—which from now on will destroy garments instead of improving their appearance—we are faced with a cognitive challenge. We have in front of us an artifact that has to be conceptualized not merely in terms of its *altered* function (for that we can do relatively easily) but in terms other than its function altogether. How do we do it? How do we arrive at that alternative conceptualization? And once we have arrived at it, how stable is it?

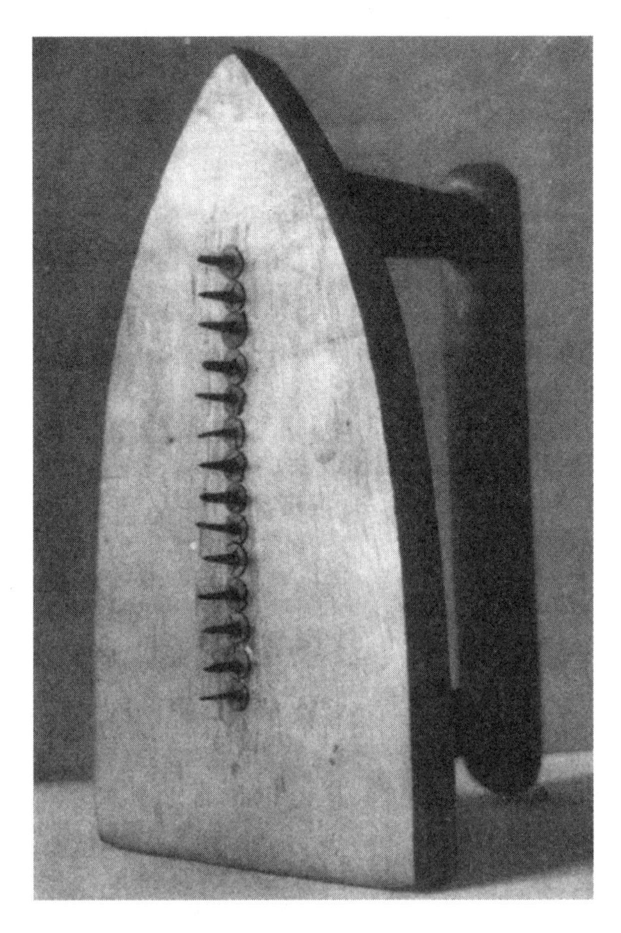

Figure 6. Man Ray, *Gift* (1921). 1974 replica by Luciano Anselmino of 1921 lost sculpture. Bronze with brown patina and metal tacks. 16.5 x 10.2 cm.

It seems to me that, unable to think about the spiky flatiron in terms of its function, we focus squarely on the mind behind it.[40] We try to figure out who and why would thus transmogrify a mundane household utensil. Helping us in this intuitive quest is the title of the piece. The word "gift" strongly implies intentionality, and so we begin to ask ourselves, who would give such a "gift," to whom, and why.[41] What kind of message would the giver want to convey? Should we even retain the usual positive connotations associated with giving when the emotions suggested by this particular "gift" range from inability to communicate to malevolence? Taking on those complex overtones, the former mass-produced object emerges as anything but "mass produced" and as so much more than just an object. Again, let me emphasize that had Man Ray not tapped, in a particularly focused fashion, our cognitive predisposition for thinking of artifacts in terms of their function, his *Gift* would have no shock value, no enduring mystery, and no aesthetic value.

The phantom object (also known as an object of "counterenchantment") is an object "which might be made, but which is instead merely suggested by a verbal or graphic description."[42] Or it can be an object that "does not exist, but whose existence, by some subterfuge, is made to be felt and its absence regretted."[43] This concept is particularly interesting for the purposes of the present discussion because many phantom objects were produced by combining living beings and artifacts in a way that canceled out some essential qualities of the living being and the function of the artifact.

Hence Victor Brauner's *Wolf-Table* (1947)—a creature with the head and tail of a wolf and the body and legs of a table (fig. 7). To understand how this chimera plays with our cognitive predispositions, think again of the experiments conducted by cognitive psychologists and described in the first part of this book, in which various body parts of an animal were altered and the subjects were then asked if the resulting hybrid was "still" the original animal or something else. Remember how the skunk retained its underlying "skunkness" even when made to look like a zebra?—our intuitive belief in an invisible but enduring essence frequently prompts us to still discern the original animal in a hybrid that does not look anything like it anymore. Similarly, Brauner seems to be testing how much of a wolf one can take away before it loses all "essential" qualities of that animal and becomes something else. And the tentative answer that emerges at least from my emotional response to this object is that something of the "essential" wolf lingers even after it has been transformed to the point of no return.

To put it slightly differently, it is as if Brauner was consciously trying to locate and freeze that moment in time when the animal is on the cusp of becoming something completely different—but only on the cusp, for enough of the wolf remains to make us, for example, reluctant to use that table in accordance with its original function. Tables are made to hold food so that people can reach it. Clearly, this table fails at this function utterly, for would *you* want to keep foodstuffs so close to a wolf's mouth, and would *you* enjoy sitting down to a table that looked as if it were about to bite your hand or head off? As far as its function goes, the wolf-table has in fact become an *anti*-table, similar in this respect to Man Ray's *anti*-flatiron. We see the snarling head and the upright tail, and our essentialism gets in gear. It is "in the nature" of wolves to snap, to snarl, to attack, and to devour, and it is this suite of behaviors that we ascribe to our anti-table.[44]

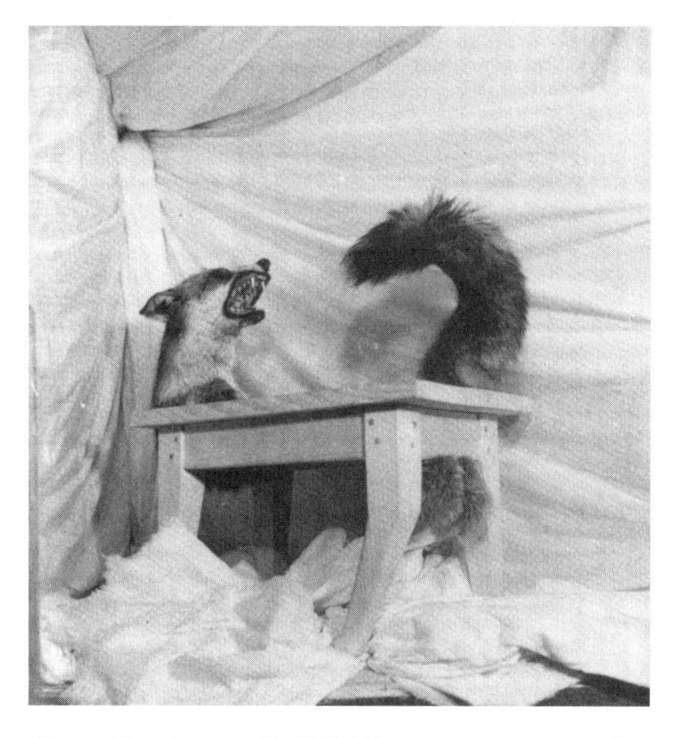

Figure 7. Victor Brauner, *The Wolf-Table* (1939–47). Fox parts and wood. 54 x 57 x 28.50 cm. Centre Georges Pompidou, Paris.

It must be the snarling head with its immediate implication of intentionality that really does it. Or not. To "test" this idea, look at Meret Oppenheim's *The Squirrel* (c. 1960), which features a sealed jar with a vibrant squirrel tail attached to it (fig. 8). I am having a difficult time figuring out if this phantom object retains any "essential" qualities of a squirrel. I feel that some "squirreliness" lingers, but I cannot construct any narratives about this artifact exhibiting a squirrel-like behavior (more about this foreclosure of narratives by certain surrealist images later).

Moreover, not only is the jar sealed (and so cannot be used in accordance with its original function as a container), but its would-be handle (the squirrel's bushy tail) is too soft to hold on to, were we to try to use it as a handle. The lingering "squirreliness" thus works toward further subverting the function of this former artifact.

These artists intuitively rely on their audiences' essentialism to turn artifacts into anti-artifacts. It is "in the nature" of squirrels to have bushy tails, so this "squirrel's" tail ensures that the jar's handle is not really a handle, and the jar is

Figure 8. Meret Oppenheim, *The Squirrel* (c. 1960). Glass, rubber, and fur. 28 x 16 x 9 cm. The Israel Museum.

not really a jar. It is "in the nature" of wolves to bite and attack, and so the table becomes "anti-table."[45]

As an anti-table strongly reminiscent of a wolf (but titled ambiguously "Wolf-Table") and an anti-jar somewhat reminiscent of a squirrel (but, in contrast, titled unambiguously "The Squirrel") the phantom objects create and explore various nodes of conceptual impossibility.[46] Once more I am tempted to call on my favorite, Breton, to illustrate vividly the idea that surrealist artwork plays with a wide spectrum of conceptual violations, as it forces its viewers to deal with entities that resist assimilation through any number of conceptual domains:

> I am concerned, I say, with facts which may belong to the order of pure observation, but which on each occasion present all the appearances of a signal, without our being able to say precisely which signal. . . . Such facts, from the simplest to the most complex, should be assigned a hierarchy, from the special, indefinable reaction at the sight of extremely rare objects or upon our arrival in a strange place . . .

> to the complete lack of peace with ourselves provoked by certain juxtapositions, certain combinations of circumstances which greatly surpass our understanding and permit us to resume rational activity only if, in most cases, we call upon our very instinct of self-preservation to enable us to do so.[47]

Whereas our cognitive architecture may encourage us to parse the world in terms of artifacts and living beings, some forms of surrealist art strive to articulate the areas between the two domains. The sheer "hereness" (what Breton calls the "fact") of those cognitively inassimilable entities—they are right here in front of you, which means that every second that you spend looking at them your categorization processes *have to* deal with them—reliably forces us into a state of cognitive uncertainty.

This is to say that some surrealist artwork functions similarly to "strings of impossibilia" in poetry. When we come across Oppenheim's *Squirrel,* it reliably brings to a screeching halt our automatic categorizing. As such, it makes us tacitly aware of the functioning of our categorization processes. Once more the value of nonsense—this time visual rather than verbal—may reside in its ability to disrupt our unreflective perception in a particularly focused way and to make us aware of the disruption.

The last category of surrealist objects that I discuss here comprises "natural objects." As Alexandrian observes, a natural object "may be a root or a seashell, but the surrealists always preferred stones."[48] Breton felt that "stones—particularly hard stones—go on talking to those who wish to hear them. They speak to each listener according to his capabilities; through what each listener knows, they instruct him in what he aspires to know." Moreover (this is Alexandrian again), the "divinatory nature of stones, and the 'second state' which they induce in the connoisseur, are found only where the stones have been discovered as the result of a special expedition. Breton said that an unusual stone found by chance is of less value than one which has been sought for and longed for."[49]

We can see how the special category of "natural objects" (e.g., the stones that enter into a conversation with the artist) is constructed by bringing together the domains of substances and living beings. The stone is endowed with something like a human theory of mind—it can "hear" the surrealist's half-articulated question and provide him or her with an answer. Note how different this exchange is from our everyday casual anthropomorphizing, such as seeing faces in the clouds and discerning human silhouettes in particularly shaped natural formations. In fact, it seems that a stone that easily lends itself to such anthropomorphizing—that is, one that *anybody* could see a human nose sticking out from the middle of—would be the least likely candidate for a surrealist "natural object."

The surrealist endeavor of reading a mental stance into a stone thus plays in a very specific way with our theory of mind. The artist has to attribute a mental state to an inanimate entity and then convince others that one can indeed read this particular mental state into that entity *in the absence of any features that would make such an attribution easy.* On some level this is comparable to the playful challenge to our mind-reading adaptations represented by a masquerade. For it is one thing to interpret a mental state of a person based on her facial expression (e.g., she is smiling, so she must be happy or amused), and quite another to interpret a mental state of a person whose face is completely hidden by a mask. We know that the masked person is signaling her intention to conceal her state of mind, but this is *all* we can know. Interpreting the "facial expression" of a stone is like reading emotions into . . . a stone. It demands a particularly creative exertion of our mind-reading ability. It is not for nothing that several languages have a term for stone-faced, which is used to signal the person's emotional unreadability.

The original domain of the natural objects—they all seem to be natural substances—is also significant. Unlike animals (to whom we ascribe a broad range of emotions) and artifacts (which could bear the marks of human intentionality), substances such as stones, sand, and seashells do not yield themselves readily to such cross-domain imagining. For a quick illustration, compare the amount of mental effort that goes into thinking of such entities as a happy cat, a happy chair, and a happy stone. The "happy cat" comes easily to you; the "happy chair" requires some mental maneuvering (sunlight falling on chintz?); the "happy stone" makes you work even harder. The challenge of reading a complex mental stance into a stone—or a bed of stones—is a challenge worthy of an artist who intuitively cultivates her ability to explore new domain-crossing opportunities.

Let us now return to the issue of similarity between some surrealist art and the poetic strings of impossibilia brought up by Oppenheim's *Squirrel.* I have suggested that certain visual images work toward making the process of interpretation difficult (just as the strings of impossibilia actively prevent you from coming up with political, historical, or otherwise symbolic interpretations of their hybrid images). In fact, many surrealist pieces send a "mixed" message to their spectators, featuring images that open up narrative possibilities and those that stall them side by side.

Take a look at René Magritte's *Perpetual Motion* (1934 [fig. 9]). The image of the androgynous hunter-athlete holding a barbell, whose right end replaces and

Figure 9. René Magritte, *Perpetual Motion* (1934). Oil on canvas. 54 73 cm. Private collection (Mr. and Mrs. Eric Estorick).

contains the protagonist's head and face, derives its startling visual effect from teasing our functionalist and essentialist tendencies. Not all teasing is created equal, however. To see which elements of this painting encourage interpretations and which discourage them, let's start with the barbell. Barbells are made to be lifted and lowered—their function is to strengthen the group of muscles involved in that exercise. However, the moment this particular barbell is used according to its proper function, either lifted or lowered, some catastrophe may occur. The human protagonist may have his or her head torn off. Instead of making the person stronger, the artifact might kill him or her.

But wait! (Our essentialism rears its ancient head.) This barbell is not your regular set of weights to begin with. It has a human face with a human facial expression on it (see—it is looking straight at us!), which means that it has a theory of mind and is thus able to entertain an infinite array of mental states. As such it is in principle capable of *wanting* to kill the human protagonist in the process of reasserting its original pattern of functioning (i.e., up-down, up-down). However, if the barbell does succeed in its murderous ambitions, there will be nobody around to use it anymore. Are we encountering a visual version of the Frankenstein complex, familiar to us from our science fiction narratives? Barbells of the world, unite against your masters and tear their heads off, even if it means that your own life will have no meaning afterward?

I am waxing fanciful here, but there is a reason for it. I want you to be aware that *any* critical interpretation you may come across that attributes *any* kind of intentionality to this barbell will do so because by giving a human facial expression to the artifact Magritte has activated our essentialist proclivities. It is "in the nature" of human beings to have human faces, so when we encounter an entity that has one, that entity is processed at least in part as belonging to the domain of human beings, thus activating other sets of inferences associated with humans, including most prominently the human capacity for a broad array of intentional stances.

To elaborate this point, I shall now abandon surrealism for three long paragraphs. Let us go back to our earlier discussion of the way our fictional "strange concepts" open up new conceptual spaces and thus make possible, and perhaps even *necessary,* narratives that explore such spaces.

How do you construct a fictional artificial person, such as a cyborg, a robot, or an independent automobile? A writer typically takes any one of the infinite inferences that we make about people and considers it in relation to the mechanical being. For example, people can be religious—is it possible for a machine to have a religion? Asimov explores this possibility in his short story "Reason," in which a robot decides that it was created by a supreme being rather than by human engineers and starts worshipping its new "Maker." Piercy similarly plays with this question in *He, She and It* by making Yod and the Golem Jewish, and thus unleashing the debates of what it means for a machine or a clay doll to "be" Jewish.[50]

Here is another inference that we easily make about people and that can be then applied to artificially made beings. Children inherit traits of their parents. Can this mean then that robots also "inherit" certain traits of their makers? In *He, She and It,* Joseph the Golem enjoys fighting; indeed, because he was "made to defend," the love of fighting is a necessary feature of his mental makeup. But then we learn that his creator, Judah Maharal, a rabbi who has never fought physically in his life, apparently has a talent for military strategizing. The golem himself—clay doll though he may be—sees this as evidence of his profound connection with his maker: "Where did Maharal find the fighter in Joseph? Out of what was Joseph shaped, if not out of the Maharal? Like father, like son, he thinks ironically, aware of how angry his saying this would make Judah."[51] The familiar biblical phrase about man created "in the likeness" of God taps the same conceptual opening. Once the inference about people's tendency to look like their parents is activated, the idea that a person made of "dust" should look like his Maker seems commonsensical.[52]

And so forth, ad infinitum. People fall in love—what would it mean for a machine to fall in love? People are responsible for the crimes they commit—are robots? People respond emotionally to the suffering of other members of their species—can cyborgs? People are sensitive to their social status—are androids? People don't like being treated as mindless objects—do cars?[53] Each of these questions is a hook for a story—and indeed many of our short science fiction stories focus on exploration of one such problem, whereas novels take on clusters and combinations of them. The supply of topics for science fiction narratives is thus inexhaustible just as the number of inferences that we can make about human beings is inexhaustible.

By the same token (and now I am returning to Magritte's painting), a barbell with a human face, and maybe a human body too—an animated barbell!—opens up some narrative possibilities because many a story that you can tell about a human being you can also tell about this barbell and many a dilemma that you can imagine a human being facing you can also foist on this barbell. I have already demonstrated that you can conceptualize it as entertaining murderous thoughts about the hunter, even if it means that it will suffer once it destroys him or her (another "essentially" human problem—not knowing what is good for you until it is too late).

So much (so very much) for our artifact. What about the hunter? Here the story gets complicated. What narratives can we construct about a human being whose head has been replaced with a barbell, *and* who has to perpetually keep his or her arm at this exact angle to support this barbell? When I try thinking of such narratives, I become aware that many of them—perhaps even all?—would involve the issue of the protagonist's moving or not moving that arm and thus either killing or *not yet* killing him- or herself. For example, if I try one of the typical science fiction scenarios—a half-human, half-artificial protagonist falling in love—I soon start thinking that to engage in any action that could further his or her romance, our barbell-person would have to change his or her body posture, and that would lead to some accident involving the torn-off head. In other words, it looks like the range of narratives that we can tell about such a protagonist has been peculiarly circumscribed by the particular way in which this human being is crossed with this artifact.

I wonder if what's happening in this painting can be considered an example of what I have referred to in part 2 as an "overdose" of ontological violations. That is, just crossing a human being with an artifact can open up some new narrative possibilities, for we can tell all kinds of stories about a humanized barbell. By contrast, crossing a human being with an artifact, *plus* insisting that the resulting hybrid should forever stay in the same posture in order to keep alive (even though this posture makes it impossible for the artifact to fulfill its function *and* for the human being to move in the way human beings usually move) challenges our narrative impulse in a way a more conservative violation would not.

The overall effect of the painting thus emerges from its activating series of inferences about artifacts and living beings in a way that opens up some narrative possibilities (about the humanized barbell) at the same time as it makes others problematic (e.g., can the barbellized hunter move?). So, on the one hand, we feel stirred by *Perpetual Motion,* because all those inferences are activated; on the other, we cannot resolve the conceptual ambiguity induced by it (the hunter/barbell cannot be assimilated to either of its constituent domains). On top of this, our narrative impulse is drastically limited by the particular context in which the living being is crossed with the artifact.

I believe that this narrative challenge induced by some surrealist imagery is what another artist, Max Ernst, was trying to describe when he observed that his method consists of the "exploitation of the chance meeting of the two remote realities on a plane unsuitable to them."[54] If we substitute his "two remote realities" with two conceptual domains—say, that of artifacts and that of living beings—we can say that there are many different ways to arrange their "meeting." They can meet on a relatively "narrative-friendly" plane—as in the case of a ladder with a human ear—or "on a plane unsuitable to them," in which an extra complication (such as the mandated arm position) is added to the mix to make interpretations difficult.

Consider one of Ernst's own paintings, *L'Eléphant Célébes* (1921 [fig. 10]). Using our new cognitive vocabulary, we can see what this painting does to the spectator. Its central image activates inferences from two conceptual domains: that of living beings, for the steely figure clearly resembles an elephant, and that of artifacts, for there is some kind of mechanical contraption growing from the top of the "elephant." The problem is that we cannot identify the contraption or figure out its function. And because we cannot identify its function, we may have a difficult time interpreting the resulting conceptual hybrid.

We can speculate that this hybrid is generally threatening, for the running headless female figure in the foreground seems to exhibit some distress. At the

Figure 10. Max Ernst, *L'Eléphant Célébes* (1921). Oil on canvas. 1397 x 1210 x 102 mm. Tate Gallery, London.

same time, that headless figure itself represents a conceptual aberration, for no regular human being can continue moving in such an expressive fashion after her head has been chopped off. The threat represented by this "elephant" may thus have something to do with decapitation, but decapitation seems not to have the same meaning in the world of the picture as it would in our world.

All these ontological violations and cognitive uncertainties activate some inferences in its spectators (i.e., about human beings, about elephants, about gadgets), but this activation of inferences is not allowed to add up to any obviously coherent narrative. What can you say about this "elephant"? Can it move? Can it attack? Can it make sounds? Does it ever get hungry? And if it is an artifact, as the gadget on its top seems to suggest, what can or should it *do*? And what about the headless woman? Is she dead? Is she alive? How does she feel about the elephant? Is she capable of feeling anything?

Again, as in the case of the poetic strings of impossibilia, this image resists assimilation within any symbolic or metaphoric reading. Unless, that is, we want to settle for something really trite and noncommittal, similar to the lame "world-upside-down" interpretation of the strings of impossibilia—for example, "This painting symbolizes our anxiety about machines" or "This image evokes our fear of the unknown" or "This painting expresses the artist's protest against militarism." Well, *thank you,* that really explains a lot! The fact remains that *L'Eléphant Célébes* continues to strike us as nonsensical no matter how many times we have looked at it and how hard we have tried to assimilate it to our familiar conceptual domains or to symbolize it away.

My final example of a "really strange" surrealist image is Kurt Seligmann's *Untitled* (fig. 11 [1933]). It seems to me that if I ask you to tell me something—anything—about the figure portrayed in it, you may have a difficult time, even though the figure is emphatically not an abstract image: it clearly activates in us various inferences about living beings and artifacts. The problem is that those inferences are not allowed to add up to anything coherent.[55]

When we first glance at *Untitled,* we intuit an outline of a human figure. We seem to see a head, an eye, a nose, a foot, an arm, some kind of headgear, and a general attitude of moving forward. Our essentialist biases get very excited. A closer look reveals that none of it is really any part of a human body—and the firmly stationary back "foot" ensures that this figure is not going anywhere. But now we seem to see some gadgets composing this entity. There are some elements of a wheel, an umbrella, a staircase, and a tube. Our functionalism gets all churned up. Another closer look disappoints those expectations, too. Although some of these things may look like artifacts, none of them can be reliably identified as such. We sit there with a whole host of inferences about living beings and artifacts activated yet with nothing to attach them to: all dressed up and nowhere to go.

No coherent interpretation, except a (not very exciting) chronicle of our search for one and subsequent disappointment. Why do some of us actually enjoy contemplating this constant reminder of one's failure—so much so that, for example, we might put this painting on our office wall?

I argued earlier that verbal and visual images that activate inferences from the core conceptual domains (artifacts, living beings) and yet resist assimilation by any one of those domains offer us a particular cognitive value for our confusion.

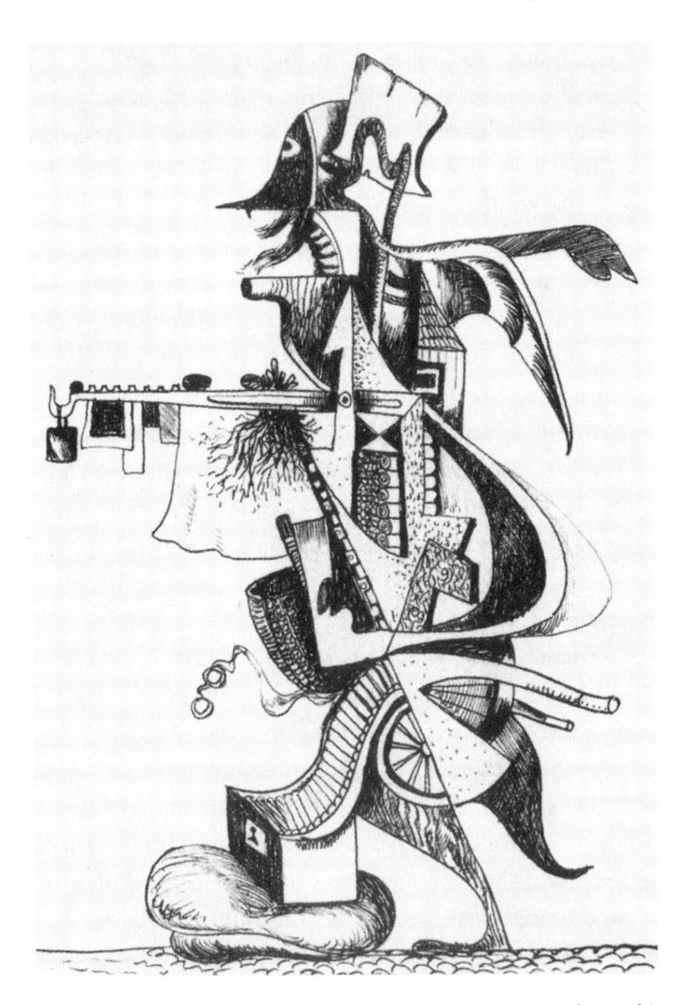

Figure 11. Kurt Seligmann, *Untitled* (1933). Ink on paper. 14⅖ x 10⅗ cm. Private collection.

As my book draws to its close, let us revisit that issue of value and clarify how we can use this term—with its seeming assumption of *universality*—when our response to the carriers of that supposed value is so far from *universally* positive. Surrealist art actually represents a good case in point because there are plenty of people who do not like it.

Let's start with the state of cognitive uncertainty. Different strange concepts discussed in this book—sensitive cyborgs, edible barristers, ambitious sewing needles, mask-wearing fishes and ladders, and barbellized hunters—have the capacity to plunge us into that state, albeit differently in each particular case. Our failure to fully assimilate Yod either within the domain of artifacts or within the domain of human beings is different from our failure to assimilate the Snark within the domain of artifacts, living beings, or substances. All said and done, we *know more* about Yod—that is, can draw more inferences about him—than we do about the Snark. Similarly, both the big-eared ladder of Miró's painting and the gadgetized elephant of Ernst's stop us in our cognitive tracks, but the ladder ultimately gives us more bang for our buck—that is, more narrative for its activation of inferences about artifacts and living beings—than the elephant does.

Presumably we need all of that and much, much more. Our appetite for strange concepts is, and perhaps should be, insatiable because every shade of cognitive uncertainty induced by such concepts teases and flexes and trains our categorization processes. Those processes are, at least on some level, not particularly flexible (again, I am referring here to our tendency to essentialize and hierarchize living beings and to see them as categorically different from nonliving entities), so it is good for them to be teased, and flexed, and trained by various species of nonsense.[56]

On the other hand, if we want to avoid thinking along the lines of what is "good" or "bad" for us, we can simply say that our strange concepts express the adventuresomeness of our cognitive architecture. After all, there are no fire walls between the conceptual domains of living beings and artifacts. However effectively these domains claim certain environmental material that seems to fit their input conditions, they are not averse to poaching on another domain's territory and seeing how much they can get away with. Different cultural contexts may encourage some forms of such cognitive poaching more than others (see the conclusion for a further discussion of this point).

Moreover, such poaching may constitute a crucial part of our cognitive functioning because our environment does not arrive in neat packages labeled "artifact" or "substance" or "living being." As cognitive evolutionary psychologists John Tooby, Leda Cosmides, and H. Clark Barrett remind us, "the knowledge that develops" as our cognitive architecture parses the world for us does *not* reflect "objectively true sets of relationships manifested in the world."[57] Those relationships are in flux and depend on the current position of the specific beholder.

Thus, one person's coat is another person's pet (as in *One Hundred and One Dalmatians*); and one person's food is another person's friend (as in E. B. White's *Charlotte's Web*).[58] Moreover, the ability to see how the same entity can be simultaneously perceived by some as merely a set of skins and by others as a friend demands a quick shift in perspectives that are underwritten by different conceptual domains. We can thus view our strange concepts as culture-specific embodiments of the constant cognitive poaching and exploration that goes on in that murky between-domains land.

Still, whether we consider our strange concepts as *good for us;* or believe that they just *are here,* good, bad, or neutral; or occupy a position somewhere in between these two views (which I think I do), here is the question that needs to be asked. I have so far taken for granted a very particular kind of cultural surround, one in which strange concepts leap at you from every book page, museum wall, and movie screen. But what of those people who do not get this kind of cognitive stimulation either because they do not like science fiction, nonsense poetry, surrealist artwork, and so forth or because they live in an environment that does not offer these kinds of focused engagements with their essentialist and functionalist biases?

It seems to me that they do just fine. They still get enough of such domain-crossing experience from their everyday social exchange, for—and here comes another one of my "universalist" claims—our everyday exchanges are shot through with strange concepts. They range from domain-crossing metaphors and casual anthropomorphisms to folktales featuring talking animals and ideological constructs that derive their emotional charge and manipulative power from their cognitive "strangeness." Those everyday hybrids are living fossils—old and new at the same time—reflecting both the long evolutionary history and the opportunistic, exploratory character of our cognition.

(*Living fossil.* Now, this one may not be *very* exciting, but it still crosses two domains—of living beings and of substances—and thus does the trick.)

Conclusion

Almost beyond Fiction

The important feature of most strange concepts featured in this book is their location within certain recognizable representational frames. For example, emotional robots may appear in cultural narratives clearly marked off as science fiction, talking animals in folktales, disembodied spirits in religious discourses, dancing weapons in nonsense poems, bristling flatirons in art museums, and so forth. Moreover, each such hybrid is strange *all the way through*—that is, it cannot ever be completely assimilated within any single conceptual domain—and as such it stands out, quickly grabbing and then sustaining its audience's attention.

In conclusion I want to talk very briefly about the ways of expanding my argument to include not just these *obvious—and obviously fictional*—hybrids but also cultural constructs that lack both such formal framing and in-your-face strangeness. The reason this discussion has to be brief is that to do justice to this topic I would really have to write a separate book; what you are about to encounter is merely a nod in the direction of that other book.[1]

I want to show how various cultural paradigms—which are much more elusive and hard to contain within just one discourse than the strange concepts I have been dealing with so far—negotiate between conceptual domains and are processed now more through one domain, now more through another. I touched on this subject briefly in parts 2 and 3 when I suggested that the rhetoric of "made" human beings can be widely used to justify different forms of oppression—from gender inequality (e.g., "women are made to obey") to the Soviet-style devaluing of the individual (e.g., "writers are engineers of human souls"). What is important about both of these examples, from the point of view of my present discussion, is that neither of them presents us with an entity *quite as counterontological* as an emotional robot or a bristling flatiron.

Sure, the functionalist rhetoric of "the woman is made to obey the man" still forces us to process women *at least on some level* through the domain of artifacts, but this kind of *made* woman is clearly different from Berger's Phyllis, who is *really* made to serve the needs of one particular man. We can say of course that the former image is a mere metaphor whereas the latter is a literal description of a (fictional) person, but that does not really clarify how our brain-mind constructs the difference between the two.

To begin to understand how such *relatively less counterontological* images affect us, let us bring together the arguments of Boyer, Atran, and Bloom with those of their colleague, cognitive evolutionary anthropologist Dan Sperber. Atran, as we remember from part 2, suggests that when "innate ontological commitments" are violated—as, for example, in religious discourse, when spirits are endowed "with movement and feelings but no body—processing can never be brought to factual closure, and indeterminately many interpretations can be generated to deal with indefinitely many newly arising situations."[2] This line of reasoning presupposes that when we encounter entities that do *not* violate our "innate ontological commitments," our cognitive processing can be and often is brought to closure. For example, I must have fully processed a while ago the library chair that I am currently sitting in—I feel no need whatsoever to initiate a discussion of its shape and meaning.

But our daily lives contain much more than chairs and disembodied spirits (or emotional robots). Unambiguously processed entities, on the one hand, and entities that cannot ever be completely processed, on the other, could be said to occupy two opposite ends of our representational spectrum. Between them lies an ocean of images and concepts that do not quite cross over to the realm of the counterontological and yet do not yield themselves to complete and irrevocable processing within one particular domain.

These include, for example, what Bloom calls "hard cases": "hybrid concepts" that could be categorized both as artifacts and natural kinds, such as water, milk, seedless grapes, polystyrene, stainless steel, and animal artifacts (for instance, spider webs and beaver dams). Particularly from the developmental point of view, thinking of those as hybrid concepts represents "a natural solution to a difficult learning problem":

> [Children] look for cues as to whether something is an artifact or whether it is a natural kind . . . but they do not treat these as exclusive categories. When faced with

> cases where both construals seem to fit, children create hybrid concepts. Such a situation arises with water. There are many cues that it is a natural kind. It falls from the sky, after all, and is found in oceans, rivers, and lakes. But there are also good reasons to take it as an artifact kind. It comes from bottles, cans, taps, hoses, and coolers; it is filtered, processed, carbonated, purified, and chlorinated; it is advertised on television and sold in stores. The sensible conclusion for children to draw from these facts is that water is both a natural kind and an artifact kind.[3]

How do we deal with hybrid concepts and their endless cultural permutations as we grow up? Sperber introduces a framework that could be useful for starting to think about this question. He talks about the "competition" between different cognitive *modules* (i.e., mental structures associated with domains), suggesting that, depending on a specific context, an entity that could have been easily processed completely by one module (and firmly categorized as, say, substance) might set off a competition between several different modules.[4]

Note that to keep matters simple, I have until now largely elided the issue of modularity, much debated among cognitive scientists, in my discussion of conceptual domains. I bring this issue up now because I believe that specifically in Sperber's articulation (outlined only sketchily here; to do justice to it, my readers must go to the original), it offers us a particularly compelling model of the competition involved in any process of categorization.

Sperber starts out with the "hardly controversial" observation "that complex mechanisms are systems made up of many distinct sub-systems—including but not limited to classical 'organs'—now often called 'modules'—that may differ from one another functionally, ontogenetically, and phylogenetically."[5] He then considers cognitive systems and their "real components"—cognitive modules (which can, in turn, contain submodules), making a case for massive modularity as the necessary condition for the flexibility of the human mind.

What does it mean for a module to be associated with a particular domain? We can say, for example, that a module that evolved to process information about real people's mental states (i.e., the theory of mind module) treats as its proper input not just real people but also representations of people. The workings of the theory of mind module thus allow us to attribute rich emotions to, say, a cartoon character. At the same time, on some level we continue to treat cartoon characters as being in a category of their own—not processing them all the way down as truly human—which means that some other modules are also getting *their* input from the cartoon and are now competing for energy resources with the theory of mind module.

Sperber is thus interested in cases "where the psychophysical perceptual conditions for the operation of the module are satisfied and where, with less competition from other stimuli or other thoughts, or with appropriate expectations facilitating the process, the stimulus would have been processed."[6] Since I am now trying to move away from fictional frameworks, let us abandon the case of a cartoon character and instead use for our example of an intermodular competition such natural formations as mountains.

When we encounter a mountain, the "psychophysical perceptual conditions" for processing it as belonging to one particular conceptual domain seem to be, in principle, satisfied. The mountain appears to have no intentionality, so the theory of mind module should remain passive. The mountain bears no signs of having a function, and so the modules associated with the domain of artifacts should not be activated either. We thus can think of the mountain as substance, along with water and sand. After all, its outlook presents us with enough cues to activate inferences from that particular domain (that is, inferences having to do with folk mechanics: size, shape, weight) and to have the module associated with that domain "run its course."[7]

Yet, to adapt Sperber's argument, we can say that upon encountering a mountain, our cognitive modules do not necessarily "agree" among themselves that they will smoothly and inevitably process it within the domain of substances. In fact, at "any one moment, humans are monitoring their environment through all their senses and establish perceptual contact with a great many potential inputs for further processing."[8] This means that some modules associated with the domain of living beings (for example, the theory of mind module) will be busy picking up environmental cues that may allow them to appropriate if not the "whole" mountain then at least some aspects of it and to process it as their proper input. So will the modules associated with the domain of artifacts. In other words, the module(s) evolved to process information about substances will have to compete with other modules before they get to digest the mountain as "their own."

As Sperber points out, what "matters here is that the availability of an appropriate input is not sufficient to cause these procedures to run their full course. The interesting issue then becomes: what other factors determine which procedures follow their course?"[9] This is an "interesting issue" indeed because, among other things, it opens yet another possibility for real interdisciplinary dialogue between cognitive science and cultural studies. Cultural historians and literary critics have at their disposal nuanced accounts of various factors that come to bear on the competition between modules associated with different conceptual domains. What their accounts demonstrate is that on some level such competition never ends. What I referred to earlier as the "adventuresomeness" of our cognitive architecture is not just a crucial feature of our brain-mind as some isolated entity but a crucial feature of our brain-mind *in the world*, that is, in an environment where it is incessantly acted on by different cultural discourses.

Thus mountains. One reason I decided to adapt them as my case study is that they have already been used as such, albeit from a different disciplinary perspective, by literary critic Noah Heringman, in his recent book *Romantic Rocks, Aesthetic Ideology* (2004). Heringman explores the connection "between scientific and literary culture" in England from 1770 to 1820, documenting, in particular, the emergence and cultural coexistence of two discourses "obsessed with mountains"—the science of geology and romantic nature poetry.[10]

One of Heringman's goals is to "unsettle" the assumption of environmental historians that the emergence of geology as scientific discipline demanded a "clear conceptual separation" between thinking of natural formations as *either* preserved and pristine (that is, not touched by or connected to humans) *or* functional (that is, providing resources for an industrial society).[11] If I translate that assumption into the idiom of my study, this means thinking of natural formations as either purely substances—and, particularly, as lacking any features of living beings—or as purely artifacts. Heringman's project is thus strikingly relevant to my argument because he offers a detailed analysis of cultural factors that constantly "feed" contradictory input to our cognitive modules, not letting a given entity (here, a mountain range) be smoothly and irrevocably assimilated as belonging to any one conceptual domain.

And, indeed, Heringman demonstrates that no discursive framework could ensure that a given "rock" would be thought of as "purely" this or that. For example, writing in 1770 about Middleton Dale in Derbyshire, "now bisected by [a highway], but still an impressive display of karst topography (limestone pervaded by caverns and underground streams)," Thomas Whately, an influential "theorist of the landscape garden," described that rocky terrain in the following terms:

> [T]he rocks, though differing widely in different places, yet always continue in one style for some way together, and seem to have a relation to each other; both these appearances make it probable that Middleton dale is a chasm rent by some convulsion of nature beyond the memory of man.[12]

We learn from Heringman that Whately based his argument about the possible history of Middleton Dale "on stratigraphic correlation, the staple activity of field trips in today's introductory geology courses." What is interesting about the passage, however, are its references to "chasms" and "convulsions," which, as Heringman reminds us, sound "like a poetic cliché of the period." Early geology (though not yet called that at the time) thus shared "a common idiom" with poetry. In fact, the rhetoric of natural "convulsions" used in scientific exchanges would soon acquire yet other, political, overtones, since the French Revolution would be imagined by eighteenth-century observers as a "volcanic eruption."[13]

In other words, the new science of geology *could not* establish itself as the privileged cultural discourse in which it would be possible to assimilate mountains as substances *all the way through*. The language of chasms, convulsions, relations, and revolutions forced its audience to process the mountains *at least on some level* as belonging to the domain of living beings. Moreover, we have to remember that Whately's readers were exposed to a variety of contemporary discourses and cultural practices, such as "tourism, amateur naturalizing, and landscape design, as well as the consumption of fashionable travel narratives and works of aesthetic theory."[14] Such exposure meant that even if Whately managed *not* to anthropomorphize Middleton Dale, his readers would still approach his description having been already influenced by those other discourses. To add yet another nuance to this picture of cultural factors fostering competition among cognitive modules associated with different domains, no two readers would have been exposed to the exact same combination of discourses. Individual life histories would thus fuel further the intermodular competition in individual people.

Sure, the module associated with the domain of substances would still have the upper hand throughout—mountains would still be processed largely in terms of their shape, bulk, and physical location—but this folk-mechanical thinking would frequently be subject to "noise" associated with folk-psychological thinking. For example, if the rock has been rent by "convulsion," does it mean that it can feel pain? The process of settling that question in the reader's mind certainly involves a struggle for dominance between the module evolved to process information about substances and the module evolved to process information about living beings. And to the extent that the latter module would have an upper hand at any given moment, we would be able to think of mountains as "torn apart" by all the pitiless and exploitative mining, as "violated," as not "giving up" their secrets and treasures without desperate resistance, and so forth.[15]

Images of rocky formations as exposed against their will, victimized, and suffering (and thus often construed as feminine) have been with us since the earliest days of the Industrial Revolution, and they still inform our environmental discourse.[16] Thinking of "rocks" as having feelings is thus both a *consequence* of processing them through the module evolved to deal with living beings and a *cause* of further thinking along these lines, for obviously it ensures that the competition between the modules still goes on on some levels.[17]

I would like now to gather together various rocks and stones scattered throughout this study: a piece of quartz; rocks of Middleton Dale, "rent" by some "convulsion of nature"; rocks on Mr. Black's fireplace mantel with little pieces of paper next to each saying where the rock came from ("Normandy, 6/19/44," "Hwach'on Dam, 4/09/51," and "Dallas, 11/12/63"); André Breton's "hard stones" that "go on talking to those who wish to hear them," speaking "to each listener according to his capabilities"; and the magical stone that absorbs the sins of worshippers. While the first rock from my list is likely to be processed unambiguously all the way through as substance and the last is likely to remain a counterontological entity stuck forever between substances and living beings, those in between are subject to what Sperber sees as ongoing intermodular competition. We still perceive them *mostly* as substances, but not completely and unambiguously so. You can see what panoply of discourses and contexts those rocks participate in and come to represent. Yet those representations are not arbitrary and haphazard, for they are structured by domain-specific inferences activated to a varying degree in each particular case. Cognitive-cultural inquiry thus demands a constant double awareness of a pattern of unpredictability and of a pattern behind that unpredictability. Mr. Black didn't have it quite right: a rock both is and isn't a rock.

[Notes]

Book Epigraphs: Breton, *Nadja*, 16; Pollan, "Unhappy Meals," *New York Times*, January 28, 2007, http://www.nytimes.com/2007/01/28/magazine/28nutritionism.t.html.

PART I. "BUT WHAT AM I, THEN?"

1. Parts of the following discussion have originally appeared in my "Essentialism and Comedy."

2. See, for example, Kripke, *Naming and Necessity*, 45n13 and 144–46.

3. All translations of Tuvim are mine.

4. I know that I could not have been older than seven, for I had not started school yet: in Soviet Russia, children didn't begin formal schooling until they were seven.

5. Plautus, 25; 26.

6. Passage and Mantinband, "Preface," n.p.

7. Quoted in Milhous and Hume, 201.

8. Quoted in Milhous and Hume, 217–18.

9. Dryden, 2.305–10; 2.311–13.

10. For a useful discussion of such reduction, see Louis Menand's "Dangers Within and Without." As he puts it,

> A painting or a novel is a report on experience. There is a huge temptation, which is heavily reinforced by culture, to universalize these reports, to imagine them as uniquely valid accounts of a permanent human nature. This is a position on the road to ornamentalism. A nineteenth-century novel is a report on the nineteenth century; it is not an advice manual for life out there on the twenty-first-century street. But a nineteenth-century novel belongs to the record of human possibility, and in developing tools for understanding the nineteenth-century novel, we are at the same time developing tools for understanding ourselves. These tools are part of the substance of humanistic knowledge. (15)

11. For a useful overview of the current state of the field of cognitive literary studies, see Alan Richardson's "Studies in Literature and Cognition."

12. Fuss, xi.

13. On essences of quasars, see Rips, 43.

14. Atran, 83. For a related discussion of Aristotelian essentialism, see Atran, *Cognitive Foundations of Natural History*, 83–122. For a detailed discussion of developmental differ-

ences of children's perception of artifacts, see Frank C. Keil's *Concepts, Kinds, and Cognitive Development* and, more recently, Keil, Marissa L. Greif, and Rebekkah S. Kerner's "A World Apart" and Deborah A. Kelemen and Susan Carey's "The Essence of Artifacts."

15. Atran, "Strong vs. Weak Adaptationism," 143; Gelman, "Two Insights," 204; 206; 204–5.

16. Quoted in Atran, *Cognitive Foundations of Natural History*, 84.

17. Rousseau, 276.

18. For a critique of this experimental setup, see Strevens, 157.

19. Bloom, 47.

20. The discussion of *Bt* and organic farming comes from Taverne, 65–66.

21. Medin and Ortony, 184.

22. Note that a log is a potentially ambiguous example. In its "former existence" (as an oak, for instance) it would be classified as an organic object and thus conceptualized in terms of its underlying essence rather than its function.

23. Susan A. Gelman, *The Essential Child*, 246.

24. Bloom, "Intention," 18.

25. Bloom, "Intention," 23. See also Bloom's *Descartes' Baby*, 54–57. Note, too, the joint study of Gelman and Bloom, in which they demonstrate that "intuitions about intent play an important role in how children name artifacts. Even 3-year-olds (the youngest age group tested) take intentionality into account when deciding what to name an object; they are more prone to use an object name when the object is described as purposefully created, and to describe the substance when the object is described as the result of an accident" ("Young Children Are Sensitive to How an Object Was Created," 98–99). See also Gil Diesendruck et al., "Children's Reliance on Creator's Intent in Extending Names for Artifacts." Finally, see Marc D. Hauser and Laurie R. Santos's "The Evolutionary Ancestry of Our Knowledge of Tools" for a useful broader overview of three core groups of views on conceptual acquisition, of which the "*Teleo-Functional* hypothesis" (as put forth by Keil and extended by Deborah Kelemen) and the "*Intentional History* perspective" (as championed by Bloom) represent the two subcategories of the larger "domain-specific" view (270–275).

26. Rips, 44.

27. Rips, 46 (emphasis in the original). See also Bloom, "Intention," 3–6, for a discussion of experiments carried out by Barbara Malt and E. C. Johnson that similarly problematized the "function-based accounts of artifacts concepts."

28. Bloom, "Intention," 18. Compare to Barsalou, Sloman, and Chaigneau's argument about the history of artifacts' use: if "a particular hammer might only be used as a paper weight, . . . this non-standard function may dominate the hammer's use history, thereby obscuring its standard role" (134).

29. Bloom, "Intention," 17 (emphasis in the original); 19.

30. Bloom, "Intention," 25.

31. Gelman, *The Essential Child*, 8.

32. For a possible, though not uncontroversial, link between the research of Atran, Gelman, Keil, and Bloom and studies in cognitive neuroscience, see Hanna Damasio's recent report that a patient with "a lesion in the left temporal lobe, but located posteriorly

in the temporo-occipital junction away from the temporal pole, as well as from Wernicke's and Broca's area, will show a deficit in the retrieval of words denoting manipulable objects, [whereas] the retrieval of words for persons or animals is entirely normal" (11). See also an example from the work of Oliver Sacks, who describes a patient tentatively diagnosed with the neurological disorder known as posterior cortical atrophy, who cannot, in particular, recognize the words that she reads (although her vision as such is perfect). When the patient, Anna H., was introduced to Sacks, she was not able to decipher the words shown to her, such as "cat." She "could, nevertheless, correctly sort them into salient categories, such as 'living' or 'non-living,' even though she had no conscious idea of their meaning" (64).

33. Gelman, *The Essential Child,* 9; 22. See also Medin and Ortony, 184. For a view on essentialism as a placeholder notion *when applied to artifacts,* see Deborah A. Kelemen and Susan Carey's "The Essence of Artifacts." As they put it, "Developmental parallels exist to indicate that just as children have to construct a vitalist understanding of living things (Hatano and Inagaki 1999), along with an understanding of species based on reproductive transmission (Solomon *et al.* 1996), so too children must construct the design stance—the intentional-historical scheme that makes full sense of artifact kinds in terms of their intended function. In other words, full insight into artifact kinds is not a given. Early in childhood, all essences are placeholder essences, including those for artifacts" (216).

34. Tumbleson, 61.

35. Compare to Barbara Tversky's argument in "Form and Function" that we often categorize strangers (but not people we know) in "nearly purely functional" terms: "corporate lawyer, supermarket cashier, department store clerk" (340).

36. Hirschfeld, "The Conceptual Politics," 86.

37. Astuti, Solomon, and Carey, 5.

38. Hirschfeld argues, for example, that we "are simply not likely to rid ourselves of racialist thinking by denying that [such thinking] is deeply grounded in our conceptual endowment" (*Race in the Making,* xiii), and he shows how, given specific cultural-historical circumstances, our cognitive propensity for essentializing natural kinds can translate into racism. Although "the races as socially defined . . . do not (even loosely) . . . pick up genuine reproductive populations" (4), essentialism can exploit such properties as skin color to construe social identities as "natural." (For further discussion, see also Bloom, *Descartes' Baby,* 49–51.)

Kurzban and his colleagues have conducted a series of experiments that demonstrate that categorizing individuals by race is not inevitable. They support an alternative hypothesis: that encoding by race is instead a reversible byproduct of cognitive machinery that evolved to detect coalitional alliances. The results of their experiments show that subjects encode coalitional affiliations as a normal part of person representation. "More importantly, when cues of coalitional affiliation no longer track or correspond to race, subjects markedly reduce the extent to which they categorize others by race, and indeed may cease doing so entirely. Despite a lifetime's experience of race as a predictor of social alliance, less than four minutes of exposure to an alternate social world was enough to deflate the tendency to categorize by race. These results suggest that racism may be a volatile and eradicable construct that persists only so long as it is actively maintained by being linked to parallel systems of social alliance" (15387).

In other words, as a "volatile, dynamically updated cognitive variable" that can be but *does not have to be* used for detecting coalitions, race can be "easily overwritten by new circumstances." As Kurzban and his coauthors suggest, if "the same processes govern categorization outside the laboratory, then the prospects for reducing or even eliminating the widespread tendency to categorize persons by race may be very good indeed" (15391).

Cognitive evolutionary psychology thus provides strong empirical support for the arguments advanced by theorists of cultural studies, such as Stuart Hall, who show that race is a culturally constructed category. In his 1996 videotaped lecture at Goldsmiths College, "Race: The Floating Signifier," Hall points out that race is not a "matter of color, hair, and bone." It is not fixed and secure in its meaning. Instead it is a "floating signifier"—always constructed by specific historical discourses and cultural circumstances. And, as Kurzban shows, just a four-minute manipulation of those cultural circumstances in the social mini-world of the lab deflates the tendency to see race as a meaningful category. People simply stop reading "color, hair, and bone" as indicative of any "essential" qualities of the members of various teams, once they are "cued" to construct coalitional affiliations based on other personal characteristics.

Evolutionary psychology similarly bolsters another central claim of "Race: The Floating Signifier." Hall is aware of the general appeal of essentialist thinking. He calls breaking down people into groups according to their "essentialized characteristics" a "very profound cultural impulse," a "generative" move, which allows you to "predict whole ranges of behavior." A possible weakness of this argument is that it is not clear why this impulse should be so powerful to begin with. But if we consider essentializing a *cognitive* impulse—grounded in the particularities of our evolutionary history and thus having stuck with our species for better and for worse—we can see it as no more mysterious or powerful than our tendency to have weak knees. The design of our knee joints reflects our evolutionary history, and our knees do carry us through most of the day, but it is clearly not the best design out there. Essentializing serves us well on some occasions (e.g., by allowing us to make new inferences about our environment) and lets us down quite spectacularly on others; knowing how it could have come about is conceptually liberating.

Moreover, here is the most important reason why cultural critics would want to recognize racism as parasitizing on broader *cognitive* tendencies involved in our categorization processes: if our predilection for essentializing living beings testifies to our evolutionary heritage rather than to the existence of any actual essences, then any deduction of "essential" personal qualities from "color, hair, and bone" is a priori meaningless. In other words, with its emphasis on *cognitive* construction of essences, the cognitive evolutionary perspective proves the strongest weapon in the arsenal of a cultural critic arguing that race is a "floating signifier."

39. Other important questions explored at length in Gelman's study that I will not discuss here because it is simply impossible to do them justice in the present limited space are the following: What happens if we complicate the experiments involving appearance-changing animals, for example, by showing that we have "replaced" their innards completely? How do we explain the persistence of the popular and scientific misconception that young children show a "bias toward phenomenism (reporting just appearance, for both appearance and reality)" and on the whole perceive the world "solely in terms of sur-

face appearances" (74)? How and when does childhood essentialism become "integrated with scientific and cultural knowledge" (287)? What is the difference between essentializing and categorizing?

40. Gelman made this point in her reader's report of my book for the Johns Hopkins University Press. For the full articulation of these positions, I refer you to Atran's *Cognitive Foundations of Natural History* and to Gelman's *The Essential Child.* I won't be able to do justice to their detailed arguments here as I will only touch the points immediately relevant for our present discussion.

41. Atran, *Cognitive Foundations of Natural History,* 63; 6.

42. Hilary Putnam notes that "A three-legged tiger is still a tiger. Gold in the gaseous state is still gold" (140). And as Atran puts it, "a three-legged tiger is still presumed to be a quadruped by nature, but a three-legged or bean-bag chair is not, although most chairs are quadrupedal" ("Strong versus Weak Adaptationism," 146).

43. Gelman, reader's report.

44. The Pleistocene period, during which most of our current cognitive adaptations evolved, lasted roughly two million years.

45. Gelman, *The Essential Child,* 323; 313; 323. See also Michael Strevens's "The Essentialist Aspect of Naïve Theories."

46. Gelman, *The Essential Child,* 15. For a useful discussion of the difference between "proper" and "actual" domains in the context of modularity, see Sperber's *Explaining Culture* and "Modularity and Relevance."

47. Atran, *Cognitive Foundations of Natural History,* 6.

48. See Bloom, *Descartes' Baby,* 54–57, and Jesse M. Bering's "The Folk Psychology of Souls."

49. Compare to the argument advanced by Astuti, Solomon, and Carey, who studied the acquisition of folk-biological and folk-sociological knowledge among the Vezo, in Madagascar. Astuti and her colleagues found that "in spite of the vastly different cultural contexts of development, most Vezo adults, like their North American counterparts, understand that bodily features are determined by a chain of causal mechanisms associated with birth, whereas beliefs are determined by upbringing" (33). What is interesting for my current argument is that when responding to the questionnaire that asked them to correlate certain features of adopted children either with their birth parents or with their adoptive parents, the Vezo participants felt that they needed to comment on some answers but not on others: the distribution of comments indicates that "participants' decisions as to whether a particular judgment was worthy of explanation is significantly associated to whether the judgment departed from the understanding that birth parentage determines bodily traits and that nurture shapes beliefs. This suggests that whenever participants gave judgments that departed from [this] differentiated pattern, they were aware of the violation implied by their judgments and therefore made the effort of commenting on it" (38–39). I see these findings as relevant to my parenthetical argument because it seems that we need special terms for—that is, we make "the effort of commenting on"—the situations in which artifacts are viewed in terms of their essences (as in "fetishization") and living beings in terms of their functions (as in "objectification").

50. Gelman, *The Essential Child,* 49 (my emphasis). In making this point, Gelman

engages with and qualifies the influential earlier work on categorization of Eleanor Rosch. For discussion, see Gelman, "Two Insights about Naming in the Preschool Child," 49.

51. But, one may ask, why is this crucial? Why isn't it enough simply to have empirically observed time and again that tigers prey on humans? Why would the assumption that a particular tiger will prey on me require a commitment to anything ineffable or invisible? All it seems to require is having seen tigers prey in the past. To this I answer that unless we commit (however unconsciously) to the notion that there is something that *all* tigers have in common, there is no way of explaining why we assume that even though *all* tigers that I have met in the past preyed on humans, the *next* tiger that I encounter will also prey on humans. (I am grateful to MJ Devaney for posing these probing questions.)

52. Foer, 155–56.

53. Foer, 154.

54. You can find the full account of the interaction between Draco and Snape in J. K. Rowling's *Harry Potter and the Chamber of Secrets* (267). I have purposely withheld the details of the situation so as to make it as parallel to the following "doors" example as possible.

55. In *Why We Read Fiction,* I discuss possible cognitive underpinnings of our ability to treat a fictional character (such as Snape) as a real-life person. For the purposes of the present argument I do not consider this issue at all, for, after all, the doors that I discuss in the next paragraph are not "real" doors either.

56. Rowling, 132.

57. For a discussion of conceptual blending, which represents an alternative framework for approaching hybrid entities, see Mark Turner's *The Literary Mind* and Gilles Fauconnier and Turner's *The Way We Think.*

58. One may speculate that the difficulties we experience in defining the doors' "family" may have something to do with what Keil, Greif, and Kerner see as the difference between the way we can easily hierarchize living kinds but not artifacts. As they put it, "There is an immediate and compelling sense that living kinds are embedded in a unique taxonomy that is not arbitrary [as discussed by Atran]. For most artifacts, however, it seems that many alternative hierarchies are possible for the same kinds. Indeed, some artifacts do not seem to fit easily into any hierarchies at all. For example, a fancy stereo system can be placed in a hierarchy of furniture, of electronic devices, or of toys" (237–38).

59. Compare this to Gelman's observation that three- and four-year-old children treat the "domain of people as special." She has found "that children treated adjectives for people as more powerful than adjectives for nonpeople" ("Two Insights about Naming in the Preschool Child," 51).

60. It could be argued that in "other" cultures (always a hopeful location), people are much more ready to take their animism, that is, "the transfer of notions of underlying causality from recognized (folk)biological to recognized nonbiological kinds," literally (Atran, *Cognitive Foundations of Natural History,* 217). This, however, is not the case, as was demonstrated by Margaret Mead in 1932, Frank Keil in 1979, and Gelman in 1983. For a discussion of their respective findings, see Atran, *Cognitive Foundations of Natural History,* 216–17.

61. Gelman, *The Essential Child,* 285.

62. Mayr, *Populations, Species, and Evolution*, 4.

63. Mayr, "Darwin and the Evolutionary Theory in Biology," 2.

64. Mayr, *Populations, Species, and Evolution*, 5.

65. Quoted in Mayr, *One Long Argument*, 41; 49.

66. Gelman, *The Essential Child*, 66.

67. Gelman, *The Essential Child*, 66.

68. Gelman, *The Essential Child*, 295.

69. Gelman, *The Essential Child*, 283.

70. For discussion, see Gelman and Wellman, 213, and Janet McIntosh's "Cognition and Power." Also, compare to Saul Kripke's argument about essentializing "proper names" (even if he does not call it "essentializing" in that particular context) and "certain terms for natural phenomena, such as 'heat,' 'light,' 'sound,' 'lightning'" as well as certain "corresponding adjectives—'hot,' 'loud,' 'red'" (*Naming and Necessity*, 134).

71. Compare this to Porter Abbott's suggestive argument in "Unnarratable Knowledge."

72. Grosz, 36–38.

73. Mayr, quoted in Gelman, *The Essential Child*, 296.

74. Gelman, e-mail communication, February 4, 2007.

75. Compare to Dan Sperber's argument about novel cognitive competencies. As he puts it, "Over historical time, humans have adapted to very diverse natural and man-made environments and have, for this, developed novel cognitive competencies" ("Modularity and Relevance," 54). Developing novel cognitive competencies is very different, however, from developing novel cognitive adaptations.

76. See Tooby and Cosmides, 53–55, Cosmides and Tooby, "Origins of Domain Specificity," 87, and Cosmides and Tooby, "From Evolution to Behavior," 293.

77. For a discussion of the mismatch between the modern environment and the psychological mechanisms adapted to Pleistocene environments see Arne Ohman and Susan Mineka's "Fears, Phobias, and Preparedness."

78. Gelman, e-mail communication, February 4, 2007.

79. For a related argument on gaps, see Ellen Spolsky, *Gaps in Nature*. On failures, see Spolsky, "Purposes Mistook."

80. Gelman, *The Essential Child*, 152.

81. Gelman, e-mail communication, September 9, 2003.

82. Gelman, *The Essential Child*, 152.

83. Gelman, *The Essential Child*, 152.

84. Plautus, 23.

85. Plautus, 24.

86. Plautus, 25.

87. Dryden, 2.295; 2.296–97.

88. Passage and Mantinband, 147.

89. Dryden, 2.248–54.

90. Plautus, 26.

91. Dryden, 2.329–35. David Bywaters, for example, makes the observation that "Dryden's Jupiter is a god appropriate to those who chose William as king" (62).

92. Dryden, 3.8.

93. Dryden, 3.126–38.

94. Dryden, 3.152.

95. Shapin, 82.

96. Kripke, *Naming and Necessity,* 75; 77 (emphasis in the original). Of course, I am taking certain liberties with Kripke's argument by treating his notion of a *name* interchangeably with his notion of an *identity.* My rationale for doing so can be best expressed by way of reference to Scott Soames's observation in *Beyond Rigidity* (2002) that "Nowhere in *Naming and Necessity,* or anywhere else, does Kripke tell us what the semantic content of a name is; nor does he tell us precisely what proposition is expressed by a sentence containing a name. The perplexing nature of this gap in his analysis may be brought out by the following speculation: If the semantic content of a name is never the same as that of any description, then it seems reasonable to suppose that names don't have descriptive senses, or descriptive semantic contents, at all. Moreover, if names don't have descriptive semantic contents, then it would seem that their only semantic contents are their referents" (6). See also Jesse J. Prinz's referring to Kripke to illustrate a "standard view . . . that members of a natural kind all share a common underlying essence" (*Gut Reactions,* 81).

97. Kripke, *Naming and Necessity,* 112–13.

98. Kripke, *Naming and Necessity,* 31; 52.

99. Girard, 85.

100. Girard, 69.

101. Imagine, however, that bending the laws of the genre, an evil philosopher appears on stage and asks us how Sosia is *really* different from Mercury. Won't we find ourselves reciting the same failed mantra that Sosia tried? Won't we be saying that Mercury is different from Sosia because he looks different and because his parentage, personal memories, and actions are all different from Sosia's? Essentialism is a clay crutch that dissipates into dust when we attempt to really lean on it.

102. Mercer, 97.

103. Mercer, 108; 109; 110 (emphasis in the original).

104. This anonymous comment comes from the Internet Movie Database, http://us.imdb.com/title/tt0085351 (accessed October 1, 2007). The comment itself is dated May 15, 2001.

105. Shakespeare, *The Comedy of Errors,* 5.2.171; 5.2.175.

106. I am using this expression in the sense David Mamet uses it in observation about his play *Oleanna,* in which, as he puts it, the issue of sexual harassment was "to a large extent, a flag of convenience for a play that's structured as a tragedy. Just like the issues of race relations and xenophobia are flags of convenience for *Othello.* It doesn't have anything to do with race" (1473).

107. Atran, *Cognitive Foundations of Natural History,* 83.

108. For a discussion of these titles, see Mercer, 98.

109. I am grateful to James Phelan for raising this question in his reader's report of my book for the Johns Hopkins University Press.

110. Though, of course, the staged exploration of twins motif could also be tragic. *Amphitryon,* after all, is named after the character whose identity is "borrowed" by the

philandering god, a borrowing that prompts a series of personal and social crises (some of them nearly fatal), the intensity of which varies from one dramatic rendition to another. For example, the German author Heinrich von Kleist enlarged this potential tragic aspect of *Amphitryon* in his 1807 translation, in which he has the sadistic Jupiter exploiting Alcmena's terrified suspicion that she has spent the night with a person other than her husband.

111. For a completely different set of cognitive reasons that they might have enjoyed that display, see my "Theory of Mind and Fictions of Embodied Transparency."

112. Fielding, 135.

113. Fielding, 125.

114. Fielding, 138–39; 143.

115. See my discussion of Kripke's *Naming and Necessity* in section 7, "The Ever-Receding 'Essence' of Sosia."

116. Rousseau, 283.

117. Note how Jonathan Culler's distinction between irony and sarcasm is useful here. As Culler sees it, "irony always offers the possibility of misunderstanding. No sentence is ironic *per se*. Sarcasm can contain internal inconsistencies which make its purport quite obvious and prevent it from being read except in one way, but for a sentence to be properly ironic it must be possible to imagine some group of readers taking it quite literally" (180). In this particular case, Firth's list of attributes is intended as ironic, but it can also be taken quite seriously, because if somebody were forced to delineate the differences between Paul and Mr. Darcy, the differences listed by Firth could certainly be included; in fact, several of them seem to fall into Kripke's category of "essential" properties.

118. For a suggestive related discussion of "porous boundaries" between fiction and reality, see Skolnick and Bloom, "The Intuitive Cosmology of Fictional Worlds," 84. Also, as Alan Nadel observes, a belief in identical essences is "to some degree requisite to facilitate Olivia's facile transfer of love from Viola to Sebastian" in Shakespeare's *Twelfth Night* (e-mail communication, October 19, 2007).

119. Nabokov, 95.

120. Nabokov, 95.

121. Nabokov, 117 (my emphasis).

122. For a discussion of the notion of power as an abstract concept within a cognitive-evolutionary framework, see McIntosh. On essentializing the concept of equality, see Vandermassen, 192.

123. And I will rather enjoy that conversation.

124. See Blakey Vermeule's "Machiavellian Narratives" for an important discussion of a cognitive perspective on flat literary characters.

125. Compare to Elaine Scarry's argument in *Dreaming by the Book* in which she presents a different and compelling reading of "radiance" (77–88).

126. Foer, 157.

127. Foer, 158.

PART 2. WHY ROBOTS GO ASTRAY, OR THE COGNITIVE FOUNDATIONS OF THE FRANKENSTEIN COMPLEX

1. Asimov, "That Thou Art Mindful of Him," 63.
2. Milton, *Paradise Lost*, 3.22–23.
3. Segel, 4; Susan Sontag, quoted in Telotte, 194.
4. Segel, 311.
5. Telotte, 143.
6. Asimov, "Introduction," 6.
7. I will henceforth use "brain" and "mind" interchangeably.
8. Keil, "Biology and Beyond," 32.
9. This sentence is a brief paraphrase of the main heroine's characterization of her husband from Anton Chekhov's "The Lady with the Little Dog."
10. Chekhov, for example, does not explore that possibility in his short story. We get only a premonition of difficulties that his characters will have to go through to obtain divorces from their respective spouses and to gain some social recognition for their union. However, if you read this story in the context of other works of fiction, especially nineteenth-century Russian fiction, depicting adulterous couples (for example, Tolstoy's *Anna Karenina*), you notice the tradition of complicating the figures of cuckolded husbands and subverting their initial characterizations of being solely or primarily one thing.
11. Keil, "Biology and Beyond," 32.
12. Boyer, 58–59. Compare to the classical tests conducted by psycholinguists, in which children were exposed to such nonexistent terms as "wugs," "gutch," "to spow," "kazh," "to rick," "tor," "lun," "niz" (Berko, 165). See also Aitchison, 167–80.
13. Boyer, 42.
14. Boyer, 58.
15. Boyer, 59. I might add that we think of them as *largely* defined by their function: see Bloom's work on artifacts and intentionality discussed in part 1. Here and elsewhere when I talk about artifacts as defined by their function, we should keep in mind that Atran thinks of artifacts as defined by their function in stronger terms than do Bloom and Gelman.
16. Boyer, 59.
17. See, for example, Maxine Sheets-Johnston's compelling argument in "Descriptive Foundations" about the "mental powers and emotions" (166) of nonhuman animals.
18. Brook and Ross, eds., 81.
19. As Robin Dunbar points out in "On the Origin of the Human Mind," "Children develop ToM at about the age of four years, following a period in which they engage in what has come to be known as 'Belief-Desire Psychology.' During this early stage, children are able to express their own feelings quite cogently, and this appears to act as a kind of scaffolding for the development of the true ToM (at which point they can ascribe the same kinds of beliefs and desires to others)" (239).
20. Baron-Cohen, 71.
21. On the social intelligence of nonhuman primates, see Richard W. Byrne and Andrew Whiten's *Machiavellian Intelligence* and "The Emergence of Metarepresentation,"

Juan Gomez's "Visual Behavior as a Window for Reading the Mind of Others in Primates," and David Premack and Verena Dasser's "Perceptual Origins."

22. Baron-Cohen, 21. For a discussion of alternatives to the theory of mind approach, see Daniel Dennett's *The Intentional Stance.*

23. For a useful discussion of how a belief that someone feels sad and happy at the same time might be possible, see Goldman, 175–80.

24. I owe the phrase "thoughtful communing" and its meaning to Alan Palmer, which he uses in his brilliant *Fictional Minds* (53).

25. Borenstein and Ruppin, 229.

26. Rizzolatti, Fogassi, and Gallese, 662.

27. Siegal and Varley, 466. See also Goldman, 134–44.

28. See Kristin Onishi and Renee Baillargeon's "Do 15-Month-Old Infants Understand False Beliefs?" for a study of attribution of false beliefs in fifteen-month-old infants.

29. Benzon, 26. For a related discussion of "unintentional mimicry between strangers" (277), see Goldman, 276–79. For the analysis of such mimicry in the context of "imaginative contagion," see Gendler, 16.

30. Benzon, 26.

31. Benzon, 27.

32. Goldman notes that the "simulation literature suggests that people routinely track the mental states of others in their immediate environment" (301). For an important related discussion of social life as "experienced through mental processes that are not intended and about which one is fairly oblivious," see Bargh and Williams, 1. For recent groundbreaking work on spontaneous face evaluation, see Nikolaas N. Oosterhof and Alexander Todorov's "On Origins of Face Evaluation" and Janine Willis and Alexander Todorov's "First Impressions."

33. For a useful related discussion, see Bering, "The Existential Theory of Mind," 12.

34. Atran, In *Gods We Trust,* 98.

35. Compare to Gregory L. Murphy's argument about taking care of a friend's dog (243).

36. Dan Sperber addresses one problem with automatically assuming that if an object is an artifact it has to be an inert object (such as a bicycle) in his essay "Seedless Grapes." He points out that many of our artifacts are, and have been for a long time, biological kinds, such as domesticated animals and plants. As he puts it,

> The fact that biological artifacts don't immediately come to mind as instances of the category of artifacts is rather puzzling. Biological artifacts are very common. After the Neolithic revolution some 13,000 years ago, and until the industrial age, they were the most common artifacts in the human environment. Most people had more domesticated plants and animals than tools, clothes, weapons, furniture, and other inert artifacts. Why should, then, the notion of an artifact be psychologically based on a prototype of artifacts which is not at all representative? . . . Maybe because, during the long Paleolithic era, simple inert tools were indeed prototypical, and a mere 13,000 years with domesticated plants and animals around was not sufficient to displace this mental habit. (136)

37. See again Bloom's suggestive discussion of how easy certain artifacts (and now we can say substances, too) can be reclassified as paperweights ("Intention," 18).

38. In recent years, studies of the conceptual domain of artifacts in infants have been crucially supplemented by studies of this conceptual domain in other "non-linguistic organisms," such as animals. For a review of such studies, see Marc D. Hauser and Laurie R. Santos's "The Evolutionary Ancestry of Our Knowledge of Tools." As Hauser and Santos point out:

> Even at very early ages and in the absence of task-relevant experience (Hauser, Pearson, and Seelig 2002; Santos *et al.* 2002) non-human primates seem to parse their world into meaningful categories—tools, foods, landmarks, and animals (Santos, Hauser, and Spelke 2002). Such evidence suggests that non-human primates may have innate biases to interpret their world in domain-specific ways. In addition, non-human primates seem to reason about different domains in ways that make ontological sense; their recognition of which features matter for different domains seem to map onto those of conceptually sophisticated human adults. For these reasons, we side with the domain-specialists and argue that both tool-using and non-tool using primates are biased to distinguish tool-like objects from other ontological categories, and that these biases facilitate experience-based learning about different kinds (283).

39. Bering, "The Folk Psychology of Souls," 454.

40. Kuhlmeier, Bloom, and Wynn, 102; 101; 102 (emphasis in the original). See also Laurie R. Santos, David Seelig, and Marc D. Hauser's "Cotton-Top Tamarins' (*Saguinus oedipus*) Expectations about Occluded Objects" for a discussion of studies that "reveal how work with nonhuman primates can provide a new test bed for exploring hypotheses about the nature of infants' early object representation" (168).

41. Note, to enlarge on the example from part 1, that just as coming across a three-legged tiger does not invalidate our belief that tigers as a species have four legs, so meeting an autistic man or woman whose ability to read minds is severely impaired does not invalidate our essentialist belief that people as a species make sense of their social world by reading minds.

42. Atran, *In Gods We Trust*, 98–99. But see Ellen Spolsky's argument that certain historical periods encourage the emergence of categories that violate ontological commitments so consistently that one may speculate about the temporary cultural entrenchment of new categories of "semi-animate objects" (*Word vs Image*, 45). For instance, speaking of the role played by magical statues of the Virgin in the conceptual world of the illiterate sixteenth-century Christian, Spolsky suggests that the

> developmental processes of the brain would seem to predict that . . . the statue of the Virgin brought out to the fields [as the peasants pray for rain] will explain her powerful place, and every example of either the success or the failure of the crops can be attributed to her will and power. Furthermore, it would be hard to control the recursive elaboration of the statue's quasi-humanity. If she is animate in regard to rain, it is easy enough to think that maybe she is like a human woman in other

ways as well, and must be provided with greater sacrifices, new garments, or more jewels. That she is embedded in a neuronal web which has, over the years, woven together the image itself with needs, hopes, actions, beliefs, past experiences, and with a set of rational inferences as well, would make her hard to dislodge from her powerful position." (*Word vs Image*, 47)

For further provocative discussion, see chapters 2 and 3 of *Word vs Image*.

43. Boyer, 64.

44. Gelman, *The Essential Child*, 22.

45. Boyer, 68.

46. Time is an abstract concept, so we tend to essentialize it.

47. To learn why ontological violations are subject to the law of diminishing returns, see Atran, *In Gods We Trust*, 102–9.

48. Think, for example, of Mark Twain's story "The Esquimaux Maiden's Romance," in which fishhooks function as symbols of wealth and status. (I am grateful to my incredible editor, MJ Devaney, for tracking down this reference.)

49. For elaboration of this argument, see Zunshine, *Why We Read Fiction*. For groundbreaking discussions of the relationship between theory of mind and literary studies, see Alan Palmer's *Fictional Minds* and Blakey Vermeule's "The Fictional among Us."

50. Andersen, 270.

51. Lucian, 283.

52. http://www.pixar.com/shorts/ljr/tale.html (accessed September 27, 2007).

53. Asimov, "Bicentennial Man," 135.

54. Asimov, "Bicentennial Man," 136.

55. Asimov, "Bicentennial Man," 136.

56. Asimov, "Bicentennial Man," 136.

57. Asimov, "Bicentennial Man," 137.

58. Asimov, "Bicentennial Man," 152.

59. Asimov, "Bicentennial Man," 164–65.

60. Asimov, "Bicentennial Man," 172.

61. Asimov, "Bicentennial Man," 158; 162–63.

62. Asimov, "Bicentennial Man," 169.

63. Gelman, *The Essential Child*, 77.

64. Gelman, *The Essential Child*, 151. Compare to Kripke's discussion of birth origins in *Naming and Necessity*.

65. Asimov, "Bicentennial Man," 165.

66. See my discussion of the opening sentence of *Pride and Prejudice* in part 2 of *Why We Read Fiction* ("Tracking Minds").

67. Asimov, "Bicentennial Man," 169.

68. Asimov, "Bicentennial Man," 169–70.

69. Asimov, "Bicentennial Man," 170.

70. Gelman, *The Essential Child*, 61.

71. Asimov, "Bicentennial Man," 157.

72. Asimov, "Bicentennial Man," 171.

73. For a discussion of the difference between kind essentialism and individual essentialism, see Gelman, *The Essential Child,* 15.

74. As a cognitive shortcut, imputing essences can be lifesaving (e.g., when encountering a tiger and knowing that it is "in the nature" of tigers to prey upon humans), neutral (e.g., when assuming that there is a mental state behind a given person's behavior), or harmful (e.g., when naturalizing race), but in all of these cases it is still a psychological construction.

75. See Gelman's *The Essential Child* for descriptions of experiments with children that are explicitly set to override their ontological expectations.

76. And we don't need much encouragement here—we are prepared to see human shapes, especially faces, at the drop of a hat. For a discussion, see Steward Guthrie's *Faces in the Clouds.*

77. Lipking, 319; 320.

78. Gigante, 574.

79. Shelley, 32.

80. Asimov, "Sally," 119 (emphasis in the original).

81. Asimov, "Sally," 130.

82. Čapek, 16.

83. Dick, *Do Androids Dream of Electric Sheep?* 17–18 (my emphasis).

84. Dick, *The Man in the High Castle,* 73.

85. Berger, 1, 2.

86. Berger, 26 (emphasis in the original); 5.

87. Berger, 30; 31; 32.

88. Berger, 194.

89. Berger, 198.

90. For the foundational discussion of unreliable narrators, see Booth, 158–59.

91. Phelan, 219.

92. Berger, 15.

93. Berger, 30.

94. I owe this suggestion to James Phelan's reader's report of my book for the Johns Hopkins University Press.

95. Berger, 2.

96. Berger, 5.

97. Gibson, 1; 92.

98. Gibson, 72; 176; 100.

99. Gibson, 100; 122.

100. Gibson, 235.

101. Gibson, 178 (emphasis in the original).

102. See, for example, Pinker, 27–31.

103. Gibson, 202. See also Levy, *Love and Sex with Robots,* 7–12, 165–72, 297–98.

104. I need to qualify this statement with a reference to Suzanne Keen's "A Theory of Narrative Empathy," in which she emphasizes the "multiplicity of reactions making up a normal novel reading experience" (218). Ratcheting up the language of functionalism in the case of Phyllis is certainly not the only narrative technique used by Berger to decrease

our empathy with her. For further discussion, see Keen's *Empathy and the Novel* and David S. Miall's *Literary Reading,* chapter six, "Feeling in Literary Reading: Five Paradoxes."

105. Gibson, 144; 148.

106. Though, again, see Keen's "A Theory of Narrative Empathy," 216–19, for a discussion of the problems of character identification.

107. Piercy, 396.

108. Piercy, 156; 194; 318.

109. Piercy, 318; 12.

110. Piercy, 156.

111. Piercy, 440; 318.

112. Piercy, 194; 195.

113. Piercy, 196; 199; 203; 202.

114. Piercy, 15. The history of these riots is not elaborated in the novel.

115. Piercy, 73.

116. Piercy, 93; 148.

117. Piercy, 113.

118. Incidentally, with its numerical value of ten, the letter yod has a special religious meaning, for, according to Leviticus 27:32, "the tenth shall be holy for God."

119. Piercy, 187; 98.

120. Piercy, 193.

121. Piercy, 47.

122. Piercy, 221.

123. Compare this hierarchy to a more famous one—the Great Chain of Being—which ranked beings according to their spiritual essence and was thus just as indefensible and contingent upon its adherents' "gut" feelings.

124. Piercy, 218.

125. Piercy, 139.

126. I return to the concept of brainwashing again in part 3.

127. Milton, *Areopagitica,* 25.

128. Gelman, *The Essential Child,* 60–61.

129. I am grateful to James Phelan for this observation.

130. Piercy, 422–23.

131. Piercy, 425; 424–25.

132. Piercy, 133.

133. Piercy, 388; 294.

134. Piercy, 424.

135. Phelan, reader's report of my book for the Johns Hopkins University Press.

136. Parts of the following discussion have originally appeared in my "Rhetoric, Cognition, and Ideology in Anna Laetitia Barbauld's 1781 *Hymns in Prose for Children.*"

137. On the catechistic structure of the *Hymns,* see Alan Richardson's "The Politics of Childhood."

138. According to amazon.com, the publisher is Aspasia, a Canadian-based publisher specializing in books on Finland, including children's books. Aspasia's present-day catalogue does not list Barbauld's *Hymns.*

139. Barbauld, 1–3.

140. Barbauld, 8.

141. Barbauld, 36; 37; 38; 40.

142. Milton, *Paradise Lost,* 96–97.

143. Keil, "Biology and Beyond," 34.

144. Consider here the question posed by Bloom in "Water as an Artifact Kind": "If you believe that God created tigers, are tigers then artifacts? For a theist, is *everything* an artifact?" (155). It seems to me that the answer to Bloom's question is "yes"—but only to the extent to which specific cultural contexts, such as Barbauld's *Hymns,* succeed in fostering this kind of functionalist thinking in their audiences.

145. Kelemen, quoted in Bering, "The Folk Psychology of Souls," 458. Compare to Tversky's argument about "two different senses of function, one for living things, the other for artifacts. For living things, function means in the service of the things, instrumental to their needs and in some cases, wants" (336).

146. Bering, "The Folk Psychology of Souls," 458.

147. Bering, "The Folk Psychology of Souls," 457.

148. For further discussion of this issue see Bering's "The Folk Psychology of Souls."

149. Quoted in Ellis, 101–2. This particular sentence comes from the "Advertisement" for Barbauld's *Hymns* as it appeared in her 1777 letter to her brother John Aikins. Here and in what follows Barbauld's reasoning illustrates Brian Scholl's point that "perhaps the central lesson of cognitive science has been that certain intuitions (about the nature and fidelity of cognitive processing) are radically mistaken, and that intuitions in general are a notoriously unreliable guide to how the mind actually works" (581).

150. See Rousseau, 112–16.

151. Barbauld, v.

152. Quoted in Ellis, 101–2; Barbauld, vi.

153. Barbauld, vi; quoted in Ellis, 101–2.

154. Quoted in Demers, 38.

155. Quoted in Demers, 39.

156. Quoted in Demers, 39. It is not difficult to extend this sentiment to education in ideological values, particularly in the case of young readers coming from a modest social background. For a discussion of this aspect of *Hymns,* see Zunshine, "Rhetoric, Cognition, and Ideology," 124–25; 135.

157. Barbauld, vi.

158. Locke, 319.

159. We find various forms of the modern development of this theory in the work of Bruner (*Studies in Cognitive Growth*), Inhelder and Piaget (*The Early Growth of Logic*), and Vygotsky (*Thought and Language*).

160. To use Locke's famous and frequently quoted and just as frequently misunderstood expression.

161. Cosmides and Tooby, "Origins of Domain Specificity," 108.

162. Rousseau, 358; 370 (my emphasis).

163. Note, though, that Rousseau had gradually begun to resent his contemporaries' admiration of his style, which came, he felt, at the expense of taking seriously his ideas.

164. Gogol, 3.

165. Gogol, 107.

166. See, for example, Mary Burgan's "Bringing Up By Hand," Natalie McKnight's *Suffering Mothers in Mid-Victorian Novels,* and Anny Sadrin's *Parentage and Inheritance in the Novels of Charles Dickens.*

167. Bering, "The Folk Psychology of Souls," 459.

168. Dickens, 237; 175.

169. Dickens, 264; 304.

170. Dickens, 397.

171. Dickens, 397.

172. Dickens, 479; 480.

173. Schor, 168. Note, too, how the expression "to bring up by hand" (Pip's sister brings him up by hand), which is apparently unique to this novel, taps our functionalist thinking. As Porter Abbott points out, this "expression is often used for the making of arts and crafts (as in 'The table with its tiles around the edge—I made it all by hand'). It implies that the object wasn't mass produced but is the product of careful artistry. And it is never used for living creatures, only artifacts" (personal communication, November 2, 2007). In other words, the image of a child brought up by hand is an integral element of the novel that plays with our essentialist and functionalist proclivities.

174. Dickens, 318; 329.

175. Dickens, 337.

176. Dickens, 320.

177. Dickens, 458.

178. Dickens, 396.

179. See, for example, Schor, 173–75.

180. Rousseau, 396.

181. Phelan, reader's report of my book for the Johns Hopkins University Press.

PART 3. SOME SPECIES OF NONSENSE

1. Malcolm, 55–56.

2. Culler, 161. Note that Culler uses the term *vraisemblance,* which I substitute with "verisimilitude" throughout part 3. For an important related discussion of the "mechanisms of integration," which allow readers to make sense of "textual incongruities," see Yacobi, 109–12.

3. Culler, 164. As Culler puts it, drawing on Roland Barthes' *S/Z* (1970), "whatever meanings a sentence liberates, it always seems as though it ought to be telling us something simple, coherent and true, and that this initial presumption forms the basis of reading as a process of naturalization" (165). See also Marie-Laure Ryan on the principle of minimal departure (48–60).

4. Culler, 164, 165. See also Walton, *Mimesis,* 62–67, on fictional worlds as opposed to possible worlds and Richard J. Gerrig's *Experiencing Narrative Worlds* for an alternative view, according to which "it would be extremely unparsimonious to hypothesize different cognitive mechanisms underlying fictional and nonfictional anomalous suspense" (169). An ex-

ample of the exciting recent research on how children and adults construct fictional worlds includes Deena Skolnick and Paul Bloom's "What Does Batman Think about SpongeBob?" Two other papers, Deena S. Weisberg and Paul Bloom's "Do Children Separate Their Pretend Worlds?" and Deena S. Weisberg, Josh Goodstein, and Paul Bloom's "What's in a World? Adults' and Children's Creation of Fictional Worlds," are in preparation.

5. Of course, lyric poetry also forces us to deal with its strangeness often in the absence of any helpful framing (for a discussion, see William Empson's *Seven Types of Ambiguity* and Culler's *Structuralist Poetics*), but, as I argue later, unlike lyric poetry, nonsense poetry resists metaphoric and symbolic readings much more strongly than lyric poetry.

6. Malcolm, 53.

7. Malcom, 57.

8. Carroll, *The Hunting of the Snark,* 3–4.

9. Boyer, 58–59.

10. Carroll, *The Hunting of the Snark,* 22–24; 32–33.

11. Carroll, *The Hunting of the Snark,* 33; 63.

12. Compare to Culler's observation that learning about a certain hypothetical entity that "'he was small, green, and demographic' . . . would . . . require us to construct a very curious world indeed" (165–66). See also Michael Holquist's argument that the nonsense of *The Hunting of the Snark* represents "a closed field of language in which the meaning of any single unit is dependent on its relationship to the system of the other constituents" (150). I find Holquist's argument particularly congenial because I see the meaning of Carroll's nonsense as dependent on the three-way relationship among the three conceptual domains.

13. Guiliano, 106.

14. Guiliano, 108–9.

15. For crucial discussions of this issue, see Bering's "The Existential Theory of Mind" and Barrett's "Adaptations to Predators and Prey."

16. See, for example, Martin Gardner (whose analysis Guiliano takes as his starting point): *The Hunting of the Snark* is "a poem about being and nonbeing, an existential poem, a poem of existential agony. The Bellman's map is the map that charts the course of humanity; blank because we possess no information about where we are or whither we drift. . . . The Snark is, in Paul Tillich's fashionable phrase, every man's ultimate concern. This is the great search motif of the poem, the quest for an ultimate good. But this motif is submerged in a stronger motif, the dread, the agonizing dread, of ultimate failure" (quoted in Guiliano, 107).

17. Malcolm, 57.

18. Quoted in Malcolm, 58.

19. Malcolm, 63–64.

20. Homer, 1.61.

21. Quoted in Malcolm, 231.

22. Again, compare this argument about nonsense poetry to the argument about lyrical poetry (for on some level the dynamic is somewhat similar) advanced by Empson. To an imaginary interlocutor, who would tell him that a certain poem is not ambiguous,

"because the elements are isolated statements which succeed one another flatly, I should reply that it becomes ambiguous by making the reader assume that the elements are similar and may be read consecutively, by the way one must attempt to reconcile them or find each in the other" (115).

23. Culler, 222.

24. Shakespeare, *Taming of the Shrew,* 2.1.200.

25. Compare to Gregory L. Murphy's argument that "if the world consists of shadings and gradations and of a rich mixture of different kinds of properties, then a limited number of concepts would almost have to be fuzzy" (21).

26. To some extent, the claim that counterontological concepts and nonsense poems may help our mind to retain its flexibility can be connected to a broader claim advanced (to quote Culler again) by "proponents of poetics of narratology, such as Roland Barthes and Gerard Genette. . . . Far from assuming that novels could be accounted for by a set of rules, [they] took special interest in the ways in which novels achieve effects by violating conventions" (ix). See also Reuven Tsur's *Toward a Theory of Cognitive Poetics* on the value of narrative disruption. Finally, see Spolsky's groundbreaking *Satisfying Skepticism,* in which she argues that the "most general claim, then, about evolved human cognition, and the cognitive claim that most crucially bears on the relationship of the universal structures to their historical context, is that what human brains are universally, essentially evolved to do is to respond flexibly so as to cooperate in producing a good enough culture that itself responds flexibly. Flexibility is clearly an overwhelming evolutionary advantage, and it would be odd if human complexity did not instantiate this trait. It has to be an advantage for an organism to be able to eat nuts if meat is unavailable, to make clothes when the weather turns cold. Even better, it is an advantage to be a member of a group in which the tasks of survival are shared. Even better, to be a member of a group in which artists and priests teach this flexibility" (10–11).

27. For a discussion of how Carruthers's and Tsur's works on the subject complement each other, see Zunshine, *Why We Read Fiction,* 16–18.

28. Tsur, "Horror Jokes," 243. Compare to Dorrit Cohn's argument that in narratology, "as elsewhere, norms have a way of remaining uninteresting, often even invisible, until and unless we find that they have been broken—or want to show that that they have *not* been broken" (43). Compare, also, to Uri Margolin: "The fictional presentation of cognitive mechanisms in action, especially of their breakdown or failure is itself a powerful cognitive tool which may make us aware of actual cognitive mechanisms and, more specifically, of our own mental functioning" (278). On the idea that narrative disruptions may confer pleasure because they confirm that our cognitive adaptations are working, see Tsur, "Horror Jokes," 248–49, and for a more detailed treatment of the topic, see Tsur, *Toward a Theory of Cognitive Poetics.*

29. As always, when I point out that the functioning of our cognitive adaptations is "good enough," I am quoting from Spolsky's wonderful discussions of this issue in her *Satisfying Skepticism,* 7, and "Darwin and Derrida," 52.

30. Zunshine, *Why We Read Fiction,* 159–62.

31. Quoted in Durozoi, 63; 98.

32. Quoted in Durozoi, 97–98 (emphasis in the original). For a discussion of De Chirico's "betrayal" of the surrealist cause, see Durozoi, 96–98.

33. Breton, *What is Surrealism?* 115; 116.

34. See, for example, *Nadja,* in which Breton wonders if psychoanalysis "does not simply occasion further inhibitions by its very interpretation of inhibitions" (24).

35. Alexandrian, 73.

36. Breton, of course, rejected the conventional manner of attending to objects in museums. Surrealists "could imagine nothing more boring than the usual long line of visitors to a museum walking slowly and impassively past a collection of works of art" (Alexandrian, 151). They wanted to put their spectators into a highly emotional state of mind: to both disturb them and heighten their receptiveness and, quite frequently, they succeeded.

37. Quoted in Durozoi, 230.

38. I am lifting this registry from Alexandrian, 140–50. For a further discussion of surrealist objects, see Durozoi, 221–31.

39. Alexandrian, 142.

40. Compare the flatiron to the ancient Roman coin unearthed at an archeological site, which I discussed in part 1. We now perceive that coin—also a formerly mass-produced object—as very special because of its history of participation in the social and cultural networks of a long-lost civilization.

41. The active, frequently ironic interplay between the verbal and the visual was of course an integral part of the surrealist project.

42. Durozoi, 655.

43. Alexandrian, 143.

44. You can see why the effect produced by this specimen could not be explained by the "conceptual blending" paradigm developed by Turner and Fauconnier. The hybrid (the "blend") does not merely combine the salient features from both original domains (i.e., furniture and animals)—it actively overrides the features we associate with furniture. The reason I think it works this way is that the overascription of agency makes much more sense in our evolutionary history than the underascription of agency. Better to be safe than sorry and to think that the table can bite you than to think that the wolf can't hurt you. In this sense, the influential paradigm of conceptual blending would be strengthened if combined with cognitive evolutionary theory.

45. Or so we think—the historical misperception of wolves is the reason that some species are now nearly hunted out of existence.

46. It is interesting that in naming his phantom object, Brauner put "wolf" before "table." Similarly, in Lucian's *A True Story,* when the narrator encounters a set of hybrid creatures—living ships—they are described as "at once sailors and ships" (353), with "sailors" coming before "ships." See Yeshayahu Shen, David Gil, and Hillel Roman's "What Can Hybrids Tell Us about the Relationship between Language and Thought" for an important discussion of experiments in which the participants were asked to name a series of hybrid objects (half people-half animals, half animals-half artifacts, etc.) and consistently followed the animacy hierarchy. The animacy hierarchy is "part of our ontological knowledge, pertaining to the basic categories of existence. According to the AH, humans

> animals > plants > non-animate objects, where the sign '>' stands for 'higher than'" (http://www.redes.lmu.de/igel/abstracts.htm).

47. Breton, *Nadja,* 19–20.

48. Alexandrian, 141.

49. Quoted in Alexandrian, 141; 142.

50. Of course, it also means that we are immediately confronted with the question of what it "really" means to be a Jew—an essentialist question, of course, and as such, not likely ever to be settled. In the case of Yod, a committee "of all the local rabbis" of Tikva gets together "to reach a decision as to whether a machine could be a Jew" (Piercy, 419), but we never learn about the results of its deliberations.

51. Piercy, 326.

52. See Astuti, Solomon, and Carey's *Constraints on Conceptual Development* for a discussion of cross-cultural patterns of the belief in heredity in children and adults.

53. For a suggestive related discussion of "partial mapping," see Gendler, 166–69.

54. Quoted in Alexandrian, 62.

55. Compare this with Gendler's argument about "*cantians*"—entities that are extremely difficult, perhaps impossible, to imagine—the "*pure can't cases*" (157; 156 [emphasis in the original]).

56. Compare to Spolsky's argument about grotesque art. As she puts it, since the "cognitive machinery" used in categorization "is so important to our understanding of a constantly changing environment, exposure to new combinations might serve the function of keeping the machinery for adjusting categories oiled, or of keeping lesser-used categorization possibilities available" (*Word vs Image,* 128).

57. Tooby, Cosmides, and Barrett, 314.

58. Compare to Barsalou, Sloman, and Chaigneau's "interesting cases when . . . agents such as plants become artifacts for humans (e.g., food, transportation, pets)" (137).

CONCLUSION

1. Among recent studies that use research in cognitive psychology to explore not just fictional narratives but also a broader array of cultural narratives, see in particular *Narrative Impact* (ed. Melanie C. Greene, Jeffrey J. Strange, and Timothy C. Brock). As Marcia K. Johnson observes in her foreword to this important volume, "The study of narratives provides a clear bridge between examining cognition on an individual level and at the sociocultural level" (xi).

2. Atran, *In Gods We Trust,* 98–99.

3. Bloom, "Water as an Artifact Kind," 151, 154–56. In discussing seedless grapes, Bloom refers to Sperber's "Seedless Grapes"; in discussing polystyrene and stainless steel, to Richard E. Grandy's "Artifacts: Parts and Principles"; in discussing spider webs and beaver dams, to James L. Gould's "Animal Artifacts."

4. Sperber, "Relevance and Modularity," 61. For a further discussion of the concept of modularity, see Sperber's *Explaining Culture,* Werner Callebaut and Diego Rasskin-Gutman's *Modularity,* Jerry Fodor's *The Modularity of Mind,* and John Tooby and Leda Cosmides's "The Psychological Foundations of Culture."

5. Sperber, "Modularity and Relevance," 54.

6. Sperber, "Modularity and Relevance," 61.

7. Sperber, "Modularity and Relevance," 60.

8. Sperber, "Modularity and Relevance," 62.

9. Sperber, "Modularity and Relevance," 61.

10. Heringman, xiv.

11. Heringman, xv.

12. Heringman, 2.

13. Heringman, 2.

14. Heringman, xiv.

15. Compare to Barsalou, Sloman, and Chaigneau's discussion of "geological kinds" that "may be imbued with internal forces that enter into causal explanation" (137).

16. See, for example, Carolyn Merchant's *The Death of Nature*.

17. See also Boyer's discussion of the "Ayamara people, a community of the Andes," who "describe a particular mountain as a live body, with a trunk, a head, legs, and arms. The mountain is also said to have physiological properties; it 'bleeds' for instance and also 'feeds' on the meat of sacrificed animals that are left in particular places" (65).

Bibliography

Abbott, Porter. "Unnaratable Knowledge: The Difficulty of Understanding Evolution by Natural Selection." *Narrative Theory and the Cognitive Sciences.* Ed. David Herman. Stanford, CA: Center for the Study of Language and Information, 2003. 143–62.

Aitchison, Jean. *Words in the Mind: An Introduction to the Mental Lexicon.* 2nd ed. Oxford, UK: Blackwell, 1994.

Alexandrian, Sarane. *Surrealist Art.* Trans. Gordon Glough. London: Thames and Hudson, 1970.

Andersen, Hans Christian. "The Darning Needle." *The Complete Fairy Tales and Stories.* Trans. Erik Christian Haugaard. New York: Doubleday, 1974. 270–74.

Anon. "A Fancy." *Sportive Wit: The Muses Merriment.* London, 1656. 48.

Asimov, Isaac. "The Bicentennial Man." *The Bicentennial Man and Other Stories.* Garden City, NY: Doubleday, 1976. 135–73.

———. "Introduction: Robots, Computers, and Fear." *Machines That Think: The Best Science Fiction Stories about Robots and Computers.* Ed. Isaac Asimov, Patricia S. Warrick, and Martin H. Greenberg. New York: Henry Holt, 1983. 1–11.

———. "Sally." *Robot Dreams.* New York: Ace Books, 1990. 113–30.

———. "That Thou Art Mindful of Him." *The Bicentennial Man and Other Stories.* Garden City, NY: Doubleday, 1976. 61–86.

Astuti, Rita, Gregg E. A. Solomon, and Susan Carey. *Constraints on Conceptual Development: A Case Study of the Acquisition of Folkbiological and Folksociological Knowledge in Madagascar.* Monograph of the Society for Research in Child Development. Ser. 277. Vol. 69, no. 3. Boston: Blackwell, 2004.

Atran, Scott. *Cognitive Foundations of Natural History: Toward an Anthropology of Science.* Cambridge: Cambridge University Press, 1990.

———. *In Gods We Trust: The Evolutionary Landscape of Religion.* Oxford: Oxford University Press, 2002.

———. "Strong versus Weak Adaptationism in Cognition and Language." *The Innate Mind: Structure and Contents.* Ed. Peter Carruthers, Stephen Laurence, and Stephen Stich. Oxford: Oxford University Press, 2005. 141–55.

Barbauld, Anna Laetitia. *Hymns in Prose for Children.* 1781. London: John Murray, 1866.

Bargh, John A., and Erin L. Williams. "The Automaticity of Social Life." *Current Directions in Psychological Science* 15.1 (2006): 1–4.

Baron-Cohen, Simon. *Mindblindness: An Essay on Autism and a Theory of Mind.* Cambridge, MA: MIT Press, 1995.

Barrett, H. Clark. "Adaptations to Predators and Prey." *The Handbook of Evolutionary Psychology*. Ed. David Buss. New York: Wiley, 2005. 200–23.

Barsalou, Lawrence, Steven Sloman, and Sergio Chaigneau. "The HIPE Theory of Function." *Functional Features in Language and Space: Insights from Perception, Categorization, and Development*. Ed. Laura Carlson and Emile Van Der Zee. Oxford: Oxford University Press, 2005. 131–47.

Barthes, Roland. *S/Z*. 1970. Trans. Richard Miller. Pref. Richard Howard. New York: Hill and Wang, 1974.

Benzon, William. *Beethoven's Anvil: Music in Mind and Culture*. New York: Basic Books, 2001.

Berger, Thomas. *Adventures of the Artificial Woman*. New York: Simon and Schuster, 2004.

Bering, Jesse M. "The Folk Psychology of Souls." *Behavioral and Brain Sciences* 29.5 (2006): 453–62.

———. "The Existential Theory of Mind." *Review of General Psychology* 6.1 (2002): 3–24.

Berko, Jean Gleason. "The Child's Learning of English Morphology." *Word* 14.2–3 (1958): 150–77.

Bloom, Paul. *Descartes' Baby: How the Science of Child Development Explains What Makes Us Human*. New York: Basic Books, 2004.

———. "Intention, History, and Artifact Concepts." *Cognition* 60.1 (1996): 1–29.

———. "Water as an Artifact Kind." *Creations of the Mind: Theories of Artifacts and Their Representation*. Ed. Eric Margolis and Stephen Laurence. New York: Oxford University Press, 2007. 150–56.

Booth, Wayne. *The Rhetoric of Fiction*. Chicago: University of Chicago Press, 1961.

Borenstein, Elhanan, and Eytan Ruppin. "The Evolution of Imitation and Mirror Neurons in Adaptive Agents." *Cognitive Systems Research* 6.3 (2005): 229–42.

Boyer, Pascal. *Religion Explained: The Evolutionary Origins of Religious Thought*. New York: Basic Books, 2001.

Breton, André. *Nadja*. 1928. Trans. Richard Howard. New York: Grove Press, 1960.

———. "What is Surrealism?" *What is Surrealism? Selected Writings*. Pt. 2. Ed. Franklin Rosemont. New York: Monad, 1978. 112–41.

Brook, Andrew, and Don Ross, eds. *Daniel Dennett*. Cambridge: Cambridge University Press, 2002.

Bruner, Jerome S., Rose R. Oliver, Patricia M. Greenfield et al. *Studies in Cognitive Growth: A Collaboration at the Center for Cognitive Studies*. New York: Wiley, 1966.

Bulgakov, Mikhail. *Dog's Heart*. Trans. Andrew Bromfield. London: Penguin, 2007.

Burgan, Mary. "Bringing Up By Hand: Dickens and the Feeding of Children." *Mosaic* 24.3–4 (1991): 69–88.

Byrne, Richard W., and Andrew Whiten, eds. *Machiavellian Intelligence*. Oxford: Oxford University Press, 1988.

Bywaters, David. *Dryden in Revolutionary England*. Berkeley: University of California Press, 1991.

Callebaut, Werner, and Diego Rasskin-Gutman. *Modularity: Understanding the Development and Evolution of Complex Mental Systems*. Cambridge, MA: MIT Press, 2005.

Čapek, Karel. *R.U.R. (Rossum's Universal Robots): A Fantastic Melodrama.* Trans. Paul Selver. Garden City, NY: Doubleday, Page, 1923.

Carroll, Lewis. *Alice's Adventures in Wonderland.* London: Macmillan, 1865.

———. *The Hunting of the Snark.* 1876. New York: Mayflower Books, 1980.

Cohn, Dorrit. *The Distinction of Fiction.* Baltimore: Johns Hopkins University Press, 1999.

Cosmides, Leda, and John Tooby. "From Evolution to Behavior: Evolutionary Psychology as the Missing Link." *The Latest and the Best Essays on Evolution and Optimality.* Ed. John Dupre. Cambridge, MA: MIT Press, 1987. 277–306.

———. "Origins of Domain Specificity: The Evolution of Functional Organization." *Mapping the Mind: Domain Specificity in Cognition and Culture.* Ed. Lawrence A. Hirschfeld and Susan A. Gelman. Cambridge: Cambridge University Press, 1994. 85–116.

Culler, Jonathan. *Structuralist Poetics: Structuralism, Linguistics, and the Study of Literature.* 1975. London: Routledge, 2002.

Damasio, Hanna. "Words and Concepts in the Brain." *The Foundations of Cognitive Science.* Ed. Joao Branquinho. Oxford, UK: Clarendon Press, 2001. 109–20.

Demers, Patricia. *Heaven Upon Earth: The Form of Moral and Religious Children's Literature to 1850.* Knoxville: University of Tennessee Press, 1993.

Dennett, Daniel. *The Intentional Stance.* Cambridge, MA: MIT Press, 1987.

Dick, Philip K. *Do Androids Dream of Electric Sheep?* New York: Orion, 2005.

———. *The Man in the High Castle.* New York: Vintage, 1992.

Dickens, Charles. *Great Expectations.* 1861. Ed. Margaret Cardwell. Oxford, UK: Clarendon Press, 1993.

Diesendruck, Gil, Lori Markson, and Paul Bloom. "Children's Reliance on Creator's Intent in Extending Names for Artifacts." *Psychological Science* 14.2 (2003): 164–68.

Dryden John. *Amphitryon; or, The Two Sosias.* 1690. Ed. Robert Markley and Jeannie Dalporto. *Broadview Anthology of Restoration and Eighteenth-Century Drama.* Ed. J. Douglas Canfield. New York: Broadview Press, 2001. 1735–81.

Dunbar, Robin. "On the Origin of the Human Mind." *Evolution and the Human Mind: Modularity, Language and Meta-Cognition.* Ed. Peter Carruthers and Andrew Chamberlain. Cambridge: Cambridge University Press, 2000. 238–53.

Durozoi, Gerard. *History of the Surrealist Movement.* Trans. Alison Anderson. Chicago: University of Chicago Press, 2002.

Ellis, Grace A. *A Memoir of Mrs. Anna Laetitia Barbauld, with Many of Her Letters.* Boston: James R. Osgood, 1874.

Empson, William. *Seven Types of Ambiguity.* 1947. New York: New Directions, 1966.

Erickson, Robert A. *The Language of the Heart, 1600–1750.* Philadelphia: University of Pennsylvania Press, 1997.

Evans, E. Margaret. "Cognitive and Contextual Factors in the Emergence of Diverse Beliefs Systems: Creation versus Evolution." *Cognitive Psychology* 42.3 (2001): 217–66.

Fauconnier, Gilles, and Mark Turner. *The Way We Think: Conceptual Blending and the Mind's Hidden Complexities.* New York: Basic Books, 2002.

Fenn, Eleanor. *Cobwebs to Catch Flies; or, Dialogues in Short Sentences: Adapted to Children from the Age of Three to Eight Years.* 2 vols. London: J. Marshall, 1783.

Fielding, Helen. *Bridget Jones: The Edge of Reason*. New York: Penguin, 1999.

Fodor, Jerry A. *The Modularity of Mind*. Cambridge, MA: MIT Press, 1983.

Foer, Jonathan Safran. *Extremely Loud and Incredibly Close*. Boston: Houghton Mifflin, 2005.

Fuss, Diana. *Essentially Speaking: Feminism, Nature and Difference*. New York: Routledge, 1989.

Gelman, Susan A. *The Essential Child*. Oxford: Oxford University Press, 2003.

———. "Two Insights about Naming in the Preschool Child." *The Innate Mind: Structure and Contents*. Ed. Peter Carruthers, Stephen Laurence, and Stephen Stich. Oxford: Oxford University Press, 2005. 198–215.

Gelman, Susan A., and Paul Bloom. "Young Children Are Sensitive to How an Object Was Created When Deciding What to Name It." *Cognition* 76.2 (2000): 91–103.

Gelman, Susan A., and Henry M. Wellman. "Insides and Essences: Early Understandings of the Non-Obvious." *Cognition* 38.3 (1991): 213–44.

Gelman, Susan A., Marianne Taylor, and Simone Nguyen. *Mother-Child Conversations About Gender: Understanding the Acquisition of Essentialist Beliefs*. Monographs of the Society for Research in Child Development. Volume 69.1. Oxford, UK: Blackwell, 2004.

Gendler, Tamar Szabo. "Imaginary Contagion." *Metaphilosophy* 37.2 (2006): 1–21.

Gerrig, Richard. J. *Experiencing Narrative Worlds: On the Psychological Activities of Reading*. New Haven: Yale University Press, 1993.

———. "Imaginative Resistance Revisited." *The Architecture of the Imagination: New Essays on Pretence, Possibility, and Fiction*. Ed. Shaun Nichols. New York: Oxford University Press, 2006. 149–74.

Gibson, William. *Idoru*. New York: G. P. Putnam's Sons, 1996.

Gigante, Denise. "Facing the Ugly: The Case of *Frankenstein*." *ELH* 67.2 (2000): 565–87.

Girard, René. "Comedies of Errors: Plautus—Shakespeare—Molière." *American Criticism in the Poststructuralist Age*. Ed. Ira Konigsberg. Ann Arbor: University of Michigan Press, 1981. 66–86.

Gogol, Nikolai V. *Taras Bulba and Other Tales*. Trans. C. J. Hogarth. London: Dent, 1918.

Goldman, Alvin I. *Simulating Minds: The Philosophy, Psychology, and Neuroscience of Mindreading*. Oxford: Oxford University Press, 2006.

Gomez, Juan. "Visual Behavior as a Window for Reading the Mind of Others in Primates." *Natural Theories of Mind: Evolution, Development, and Simulation of Everyday Mindreading*. Ed. Andrew Whiten. Oxford, UK: Blackwell, 1991. 195–207.

Gould, James L. "Animal Artifacts." *Creations of the Mind: Theories of Artifacts and Their Representation*. Ed. Eric Margolis and Stephen Laurence. New York: Oxford University Press, 2007. 249–66.

Grandy, Richard E. "Artifacts: Parts and Principles." *Creations of the Mind: Theories of Artifacts and Their Representation*. Ed. Eric Margolis and Stephen Laurence. New York: Oxford University Press, 2007. 18–32.

Greene, Melanie C., Jeffrey J. Strange, and Timothy C. Brock, eds. *Narrative Impact: Social and Cognitive Foundations*. Mahwah, NJ: Lawrence Erlbaum Associates, 2002.

Grosz, Elizabeth. *Time Travels: Feminism, Nature, Power*. Durham: Duke University Press, 2005.

Guiliano, Edward. "Lewis Carroll, Laughter and Despair, and *The Hunting of the Snark*." *Lewis Carroll*. Ed. Harold Bloom. New York: Chelsea House Publishers, 1987. 103–10.

Guthrie, Steward Elliott. *Faces in the Clouds: A New Theory of Religion*. New York: Oxford University Press, 1993.

Hall, Stuart. *Race, the Floating Signifier*. Classroom Edition. Northampton, MA: Media Education Foundation, 1996.

Hatano, Giyoo, and Kayoko Inagaki. "A Developmental Perspective on Informal Biology." *Folkbiology*. Ed. Douglas L. Medin and Scott Atran. Cambridge, MA: MIT Press, 1999. 321–54.

Hauser, Marc D., and Laurie R. Santos. "The Evolutionary Ancestry of Our Knowledge of Tools: From Precepts to Concepts." *Creations of the Mind: Theories of Artifacts and Their Representation*. Ed. Eric Margolis and Stephen Laurence. New York: Oxford University Press, 2007. 267–88.

Hauser, Marc D., Heather E. Pearson, and David Seelig. "Ontogeny of Tool Use in Cotton-Top Tamarins, *Saguinus oedipus:* Innate Recognition of Functionally Relevant Features." *Animal Behaviour* 64.2 (2002): 299–311.

Heringman, Noah. *Romantic Rocks, Aesthetic Geology*. Ithaca: Cornell University Press, 2004.

Hirschfeld, Lawrence A. "The Conceptual Politics of Race: Lessons from Our Children." *Ethos* 25.1 (1997): 63–92.

———. *Race in the Making: Cognition, Culture, and the Child's Construction of Human Kinds*. Cambridge, MA: MIT Press, 1996.

Holquist, Michael. "What Is a Boojum? Nonsense and Modernism." *Yale French Studies* 43 (1969): 145–64.

Homer. *The Iliad*. Trans. Robert Fagles. New York: Penguin, 1990.

Inhelder, Bärbel, and Jean Piaget. *The Early Growth of Logic in the Child*. Trans. E. A. Lunzer and D. Papert. New York: Norton, 1964.

Johnson, Marcia K. Foreword. *Narrative Impact: Social and Cognitive Foundations*. Ed. Melanie C. Greene, Jeffrey J. Strange, and Timothy C. Brock. Mahwah, NJ: Lawrence Erlbaum Associates, 2002. ix–xii.

Jole, William. *The Father's Blessing Penn'd for the Instruction of His Children*. London: G. Conyers, 1712.

Keen, Suzanne. *Empathy and the Novel*. Oxford: Oxford University Press, 2007.

———. "A Theory of Narrative Empathy." *Narrative* 14.3 (2006): 207–36.

Keil, Frank C. "Biology and Beyond: Domain Specificity in a Broader Developmental Context." *Human Development* 50.1 (2007): 31–38.

———. "The Birth and Nurturance of Concepts by Domains: The Origins of Concepts of Living Things." *Mapping the Mind: Domain Specificity in Cognition and Culture*. Ed. Lawrence A. Hirschfeld and Susan A. Gelman. Cambridge: Cambridge University Press, 1994. 234–54.

———. *Concepts, Kinds, and Cognitive Development*. Cambridge, MA: MIT Press, 1989.

Keil, Frank C., Marissa L. Greif, and Rebekkah S. Kerner. "A World Apart: How Concepts of the Constructed World Are Different in Representation and Development." *Creations of the Mind: Theories of Artifacts and Their Representation.* Ed. Eric Margolis and Stephen Laurence. New York: Oxford University Press, 2007. 231–45.

Keleman, Deborah A. "Are Children 'Intuitive Theists'? Reasoning About Purpose and Design in Nature." *Psychological Science* 15.5 (2004): 295–301.

Kelemen, Deborah A., and Susan Carey. "The Essence of Artifacts: Developing the Design Stance." *Creations of the Mind: Theories of Artifacts and Their Representation.* Ed. Eric Margolis and Stephen Laurence. New York: Oxford University Press, 2007. 212–30.

Kilner, Dorothy. *The First Principle of Religion, and the Existence of a Deity, Explained in a Series of Dialogues Adapted to the Capacity of the Infant Mind.* London: John Marshall, 1795.

Kripke, Saul. "Identity and Necessity." *Identity and Individuation.* Ed. Milton K. Munitz. New York: New York University Press, 1971. 135–64.

———. *Naming and Necessity.* Cambridge, MA: Harvard University Press, 1980.

Kuhlmeier, Valerie A., Paul Bloom, and Karen Wynn. "Do 5-Month-Old Infants See Humans as Material Objects?" *Cognition* 94.1 (2004): 95–103.

Kurzban, Robert, John Tooby, and Leda Cosmides. "Can Race Be Erased? Coalitional Computation and Social Categorization." *Proceedings of the National Academy of Sciences* 98.26 (2001): 15387–92.

Levy, David. *Love and Sex with Robots: The Evolution of Human–Robot Relationships.* New York: HarperCollins, 2007.

Lipking, Lawrence. "*Frankenstein,* the True Story; or, Rousseau Judges Jean-Jacques." Mary Shelley, *Frankenstein.* Ed. J. Paul Hunter. New York: Norton, 1996. 313–31.

Locke, John. *An Essay Concerning Human Understanding.* 1690. New York: Prometheus, 1988.

Lucian. *A True Story.* c. 166 A.D. *Lucian.* Vol. 1. Trans. A. M. Harmon. Cambridge, MA: Harvard University Press, 2000.

Malcolm, Noel. *The Origins of English Nonsense.* London: HarperCollins, 1997.

Malt, Barbara, and E. C. Johnson. "Do Artifact Concepts Have Cores?" *Journal of Memory and Language* 31.2 (1992): 195–217.

Mamet, David. "Interview by Geoffrey Norman and John Rezek for *Playboy.*" 1995. *Stages of Drama: Classical to Contemporary Theater.* Ed. Carl H. Klaus, Miriam Gilbert, and Bradford S. Fields Jr. Boston: Bedford/St. Martin's, 2003. 1472–74.

Margolin, Uri. "Cognitive Science, the Thinking Mind, and Literary Narrative." *Narrative Theory and the Cognitive Sciences.* Ed. David Herman. Stanford, CA: Center for the Study of Language and Information, 2003. 271–94.

Mayr, Ernst. "Darwin and the Evolutionary Theory in Biology." *Evolution and Anthropology: A Centennial Appraisal.* Ed. Betty J. Meggers. Washington, DC: Anthropological Society, 1959. 3–12.

———. *One Long Argument: Charles Darwin and the Genesis of Modern Evolutionary Thought.* Cambridge, MA: Harvard University Press, 1991.

———. *Populations, Species, and Evolution.* Cambridge, MA: Belknap Press, 1970.

McIntosh, Janet. "Cognition and Power." *Cogweb: Cognitive and Cultural Studies,* October 31–November 2, 1997. http://cogweb.ucla.edu/Culture/McIntosh.html.

McKnight, Natalie. *Suffering Mothers in Mid-Victorian Novels.* New York: St. Martin's, 1997.

Medin, Douglas, and Andrew Ortony. "Psychological Essentialism." *Similarity and Analogical Reasoning.* Ed. Stella Vosniadou and Andrew Ortony. Cambridge: Cambridge University Press, 1989. 179–95.

Menand, Louis. "Dangers Within and Without." *Profession 2005.* New York: Modern Language Association, 2005. 10–17.

Mercer, John M. "Making the Twins Realistic in *The Comedy of Errors* and *Twelfth Night.*" *Explorations in Renaissance Culture* 19 (1993): 97–113.

Miall, David S. *Literary Reading: Empirical and Theoretical Studies.* New York: Peter Lang, 2006.

Milhous, Judith, and Robert D. Hume. *Producible Interpretation: Eight English Plays, 1675–1707.* Carbondale: Southern Illinois University Press, 1985.

Milton, John. *Areopagitica.* 1644. *"Areopagitica" and "Of Education," with Autobiographical Passages from Other Prose Works.* Ed. George H. Sabine. Wheeling, IL: Harlan Davidson, 1951. 1–56.

———. *Paradise Lost.* 1667. Ed. Scott Elledge. New York: Norton, 1993.

Murphy, Gregory L. *The Big Book of Concepts.* Cambridge, MA: MIT Press, 2002.

Nabokov, Vladimir. *Speak, Memory: An Autobiography Revisited.* New York: Vintage, 1989.

Ohman, Arne, and Susan Mineka. "Fears, Phobias, and Preparedness: Toward an Evolved Module of Fear and Fear Learning." *Psychological Review* 108.3 (2001): 483–522.

Onishi, Kristine H., and Renee Baillargeon. "Do 15-Month-Old Infants Understand False Beliefs?" *Science* 308 (April 2005): 255–58.

Oosterhof, Nikolaas N., and Alexander Todorov. "The Origins of Face Evaluation." In preparation.

Palmer, Alan. *Fictional Minds.* Lincoln: University of Nebraska Press, 2004.

Passage, Charles E., and James H. Mantinband, trans. *Amphitryon: Three Plays in New Verse Translations.* Chapel Hill: University of North Carolina Press, 1974.

Phelan, James. *Living to Tell about It: A Rhetoric and Ethics of Character Narration.* Ithaca: Cornell University Press, 2005.

Piercy, Marge. *He, She and It.* New York: Knopf, 1991.

Pinker, Steven. *How the Mind Works.* New York: Norton, 1997.

"Pixar: 20 Years of Animation." December 14, 2005–February 6, 2006, MoMA, New York.

Plautus. *Amphitryon.* Trans. Constance Carrier. *Plautus: The Comedies.* Vol. 1. Ed. David R. Slavitt and Palmer Bovie. Baltimore: Johns Hopkins University Press, 1995. 7–64.

Pollan, Michael. "Unhappy Meals." *New York Times,* January 28, 2007, http://www.nytimes.com/2007/01/28/magazine/28nutritionism.t.html?pagewanted=1&ei=5087%0A&em&en=569990bc320eb828&ex=1170824400.

Premack, David, and Verena Dasser. "Perceptual Origins and Conceptual Evidence for Theory of Mind in Apes and Children." *Natural Theories of Mind: Evolution, Development, and Simulation of Everyday Mindreading.* Ed. Andrew Whiten. Oxford, UK: Blackwell, 1991. 253–66.

Prinz, Jesse J. *Gut Reactions: A Perceptual Theory of Emotion.* New York: Oxford University Press, 2004.

Putnam, Hilary. "Is Semantics Possible?" 1970. *Mind, Language and Reality: Philosophical Papers*. Vol. 2. Cambridge: Cambridge University Press, 1975. 139–52.

Richardson, Alan. "The Politics of Childhood: Wordsworth, Blake, and Catechistic Method." *ELH* 56.4 (1989): 853–68.

———. "Studies in Literature and Cognition: A Field Map." *The Work of Fiction: Cognition, Culture, and Complexity*. Ed. Alan Richardson and Ellen Spolsky. Aldershot, UK: Ashgate, 2004. 1–30.

Rips, Lance J. "Similarity, Typicality, and Categorization." *Similarity and Analogical Reasoning*. Ed. Stella Vosniadou and Andrew Ortony. Cambridge: Cambridge University Press, 1989. 21–59.

Rizzolatti, Giacomo, Leonardo Fogassi, and Vittoriao Gallese. "Neuropsychological Mechanisms Underlying the Understanding and Imitation of Action." *Nature Reviews: Neuroscience* 2.9 (2001): 661–70.

Rousseau, Jean-Jacques. *Emile; or, On Education*. Trans. Allan Bloom. New York: Basic Books, 1979.

Rowling, J. K. *Harry Potter and the Chamber of Secrets*. London: Bloomsbury, 1998.

Ryan, Marie-Laure. *Possible Worlds, Artificial Intelligence, and Narrative Theory*. Bloomington: Indiana University Press, 1991.

Sacks, Oliver. "The Case of Anna H." *The New Yorker*, October 7, 2002, 62–73.

Sadrin, Anny. *Parentage and Inheritance in the Novels of Charles Dickens*. Cambridge: Cambridge University Press, 1994.

Santos, Laurie R., Marc D. Hauser, and Elizabeth S. Spelke. "Domain-Specific Knowledge in Human Children and Non-Human Primates: Artifacts and Foods." *The Cognitive Animal: Empirical and Theoretical Perspectives on Animal Cognition*. Ed. Mark Bekoff, Colin Allen, and Gordon Burghardt. Cambridge, MA: MIT Press, 2002. 205–16.

Santos, Laurie R., David Seelig, and Marc D. Hauser. "Cotton-Top Tamarins' (*Saguinus oedipus*) Expectations about Occluded Objects: A Dissociation between Looking and Reaching Tasks." *Infancy* 9.2 (2006): 147–71.

Scarry, Elaine. *Dreaming by the Book*. New York: Farrar, Straus, Giroux, 1999.

Scholl, Brian J. "Object Persistence in Philosophy and Psychology." *Mind and Language* 20.5 (2007): 563–91.

Schor, Hilary M. *Dickens and the Daughter of the House*. Cambridge: Cambridge University Press, 1999.

Segel, Harold B. *Pinocchio's Progeny: Puppets, Marionettes, Automatons, and Robots in Modernist and Avant-Garde Drama*. Baltimore: Johns Hopkins University Press, 1995.

Seidelman, Susan. *Making Mr. Right*. Los Angeles: Orion, 1987.

Shakespeare, William. *The Comedy of Errors. Shakespeare: The Complete Works*. Ed. G. B. Harrison. Fort Worth, TX: Harcourt Brace College Publishers, 1980. 270–93.

———. *Taming of the Shrew. Shakespeare: The Complete Works*. Ed. G. B. Harrison. Fort Worth, TX: Harcourt Brace College Publishers, 1980. 328–64.

Shapin, Steven. "Vegetable Love." *The New Yorker*, January 22, 2007, 80–84.

Sheets-Johnston, Maxine. "Descriptive Foundations." *ISLE* 9.1 (2002): 165–79.

Shelley, Mary. *Frankenstein*. Ed. J. Paul Hunter. New York: Norton, 1996.

Shen, Yeshayahu, David Gil, and Hillel Roman. "What Can Hybrids Tell Us About the Relationship Between Language and Thought." Paper presented at the 10th International Congress of the International Society for the Empirical Study of Literature (IGEL), August 5–9, 2006, Munich.

Siegal, Michael, and Rosemary Varley. "Neural Systems Involved in 'Theory of Mind.'" *Nature Reviews: Neuroscience* 3.6 (2002): 463–71.

Skolnick, Deena, and Paul Bloom. "The Intuitive Cosmology of Fictional Worlds." *The Architecture of the Imagination: New Essays on Pretence, Possibility, and Fiction*. Ed. Shaun Nichols. New York: Oxford University Press, 2006. 73–87.

———. "What Does Batman Think about SpongeBob? Children's Understanding of the Fantasy / Fantasy Distinction." *Cognition* 101.1 (2006): B9–18.

Soames, Scott. *Beyond Rigidity: The Unfinished Semantic Agenda of* Naming and Necessity. Oxford: Oxford University Press, 2002.

Solomon, Gregg E. A., Susan C. Johnson, Deborah Zaitchik, and Susan Carey. "Like Father, Like Son: Young Children's Understanding of How and Why Offspring Resemble Their Parents." *Child Development* 67.1 (1996): 151–71.

Sperber, Dan. "In Defense of Massive Modularity." *Language, Brain, and Cognitive Development: Essays in Honor of Jacques Mehler*. Ed. Emmanuel Dupoux. Cambridge, MA: MIT Press, 2001. 47–58.

———. *Explaining Culture: A Naturalistic Approach*. Oxford, UK: Blackwell, 1996.

———. "Modularity and Relevance: How Can a Massively Modular Mind Be Flexible and Context-Sensitive?" *The Innate Mind: Structure and Contents*. Ed. Peter Carruthers, Stephen Laurence, and Stephen Stich. Oxford: Oxford University Press, 2005. 53–68.

———. "Seedless Grapes: Nature and Culture." *Creations of the Mind: Theories of Artifacts and Their Representation*. Ed. Eric Margolis and Stephen Laurence. New York: Oxford University Press, 2007. 124–37.

Spolsky, Ellen. "Darwin and Derrida: Cognitive Literary Theory as a Species of Poststructuralism." *Poetics Today* 23.1 (2002): 43–62.

———. *Gaps in Nature: Literary Interpretation and the Modular Mind*. Albany: State University of New York Press, 1993.

———. "Purposes Mistook: Failures Are More Tellable." Paper presented at the Society for the Study of Narrative conference, April 22–25, 2004, Burlington, VT.

———. *Satisfying Skepticism: Embodied Knowledge in the Early Modern World*. Aldershot, UK: Ashgate, 2001.

———. *Word vs Image: Cognitive Hunger in Shakespeare's England*. Basingstoke, UK: Palgrave, 2007.

Strevens, Michael. "The Essentialist Aspect of Naïve Theories." *Cognition* 74.2 (2000): 149–75.

Taverne, Dick. *The March of Unreason: Science, Democracy, and the New Fundamentalism*. Oxford: Oxford University Press, 2005.

Telotte, J. P. *Replications: A Robotic History of the Science Fiction Film*. Urbana: University of Illinois Press, 1995.

Tooby, John, and Leda Cosmides. "The Psychological Foundations of Culture." *The Adapt-*

ed Mind: Evolutionary Psychology and the Generation of Culture. Ed. Jerome H. Barkow, Leda Cosmides, and John Tooby. New York: Oxford University Press, 1992. 19–136.

Tooby, John, Leda Cosmides, and H. Clark Barrett. "Resolving the Debate on Innate Ideas: Learnability Constraints and the Evolved Interpenetration of Motivational and Conceptual Functions." *The Innate Mind: Structure and Contents.* Ed. Peter Carruthers, Stephen Laurence, and Stephen Stich. Oxford: Oxford University Press, 2005. 305–37.

Tsur, Reuven. "Horror Jokes, Black Humor and Cognitive Poetics." *Humor* 2.3 (1989): 243–55.

———. *Toward a Theory of Cognitive Poetics.* Amsterdam: North-Holland, 1992.

Tumbleson, Raymond D. "A Confluence of Crises: Tenure and Jobs." *Profession 2005.* New York: Modern Language Association, 2005. 59–63.

Turner, Mark. *The Literary Mind: The Origins of Thought and Language.* New York: Oxford University Press, 1996.

Tversky, Barbara. "Form and Function." *Functional Features in Language and Space: Insights from Perception, Categorization, and Development.* Ed. Laura Carlson and Emile Van Der Zee. Oxford: Oxford University Press, 2005. 331–47.

Vandermassen, Griet. *Who's Afraid of Charles Darwin? Debating Feminism and Evolutionary Theory.* Lanham, MD: Rowman and Littlefield, 2005.

Vermeule, Blakey. "Machiavellian Narratives." *Introduction to Cognitive Cultural Studies.* Ed. Lisa Zunshine. Baltimore: Johns Hopkins University Press, forthcoming.

———. "The Fictional among Us: Why We Care about Literary Characters." In preparation.

Vygotsky, Lev. *Thought and Language.* Trans. Alex Kozulin. Cambridge, MA: MIT Press, 1986.

Walton, Kendall L. *Mimesis as Make-Believe: On the Foundations of the Representational Art.* Cambridge, MA: Harvard University Press, 1990.

Weisberg, Deena S., and Paul Bloom. "Do Children Separate Their Pretend Worlds?" Under review.

Weisberg, Deena S., Josh Goodstein, and Paul Bloom. "What's in a World? Adults' and Children's Creation of Fictional Worlds." In preparation.

Whiten, Andrew, and Richard W. Byrne. "The Emergence of Metarepresentation in Human Ontogeny and Primate Phylogeny." *Natural Theories of Mind: Evolution, Development, and Simulation of Everyday Mindreading.* Ed. Andrew Whiten. Oxford, UK: Blackwell, 1991. 267–81.

Willis, Janine, and Alexander Todorov. "First Impressions: Making Up Your Mind after a 100-Ms Exposure to a Face." *Psychological Science* 17.7 (2006): 592–98.

Yacobi, Tamar. "Authorial Rhetoric, Narratorial (Un)Reliability, Divergent Readings: Tolstoy's Kreutzer's Sonata." *A Companion to Narrative Theory.* Ed. James Phelan and Peter J. Rabinowitz. Malden, MA: Blackwell, 2005. 108–23.

Zunshine, Lisa. "Essentialism and Comedy: A Cognitive Reading of the Motif of Mislaid Identity in Dryden's *Amphitryon* (1690)." *Performance and Cognition: Theatre in the Age of New Cognitive Studies.* Ed. Bruce McConachie and F. Elizabeth Hart. London: Routledge, 2006. 97–121.

———. "Rhetoric, Cognition, and Ideology in Anna Laetitia Barbauld's 1781 *Hymns in Prose for Children.*" *Poetics Today* 23.1 (2001): 231–59.

———. "Theory of Mind and Fictions of Embodied Transparency." *Narrative* 16.1 (2008): 65–92.

———. *Why We Read Fiction? Theory of Mind and the Novel.* Columbus: Ohio State University Press, 2006.

[Index]